Indoor Infrared Optical Wireless Communications

Indoor Infrared Optical Wireless Communications

Systems and Integration

Ke Wang

CRC Press

Taylor & Francis Group

Boca Raton London New York

CRC Press is an imprint of the
Taylor & Francis Group, an **informa** business

CRC Press
Taylor & Francis Group
6000 Broken Sound Parkway NW, Suite 300
Boca Raton, FL 33487-2742

First issued in paperback 2023

© 2020 by Taylor & Francis Group, LLC
CRC Press is an imprint of Taylor & Francis Group, an Informa business

No claim to original U.S. Government works

ISBN 13: 978-0-367-25424-7 (hbk)
ISBN-13: 978-1-032-65455-3 (pbk)
ISBN 13: 978-1-00-300040-2 (ebk)

DOI: 10.1201/9781003000402

Contents

Preface

We have witnessed the rapid development of wireless communications in the past couple of decades, and currently 4G (and upcoming 5G) and WiFi technologies are surrounding us to provide high-speed wireless connections. The availability of high-speed wireless networks is changing our daily lives. For example, commuters can now watch streaming videos, check emails, or play online games on buses or trains, and students can access digital library collections and professional software everywhere on campus. With the emerging speed-hungry applications, such as ultra-high-definition video-on-demand, virtual reality, and augmented reality, even higher wireless communication speed is expected by end users. Compared with outdoor scenarios, the wireless speed requirement in indoor environments is typically higher, especially in working and living spaces.

To satisfy the ever-growing wireless communication speed requirement in indoor environments, the optical wireless technology has been explored, where the data is carried by the lightwave and transmitted through the wireless link. One key driver for investigating optical wireless communications is the availability of almost unlimited and unregulated bandwidth, which theoretically enables ultra-high-speed wireless connections. Another key advantage of the optical wireless technology is the immunity to traditional electromagnetic interference (EMI), which becomes more attractive due to the co-existence of a large number of wireless systems nowadays.

Indoor optical wireless communications can explore both the visible light range and the near-infrared wavelength range. The visible light communication (VLC) technology becomes increasingly attractive due to the wide deployment of LEDs in the lighting infrastructure, since the LEDs also serve as the transmitters in VLC. On the other hand, the near-infrared optical wireless communication technology can provide higher modulation bandwidth and wireless channel bandwidth, and hence, it is also highly attractive in the future when higher speed is required. This book focuses on near-infrared indoor optical wireless communication technologies and systems, to provide an introduction of this exciting and rapidly developing field. Specifically, this book starts from the basic theories and models, and then discusses various advanced technologies and recent developments.

Although the near-infrared indoor optical wireless technology has seen rapid development during the recent years, there are still a number of challenges to be further studied. For example, the multi-user access principle needs further development and optimization, since traditional principles lack efficiency due to the ultra-high data rate and the physical limitations in optical wireless systems. Another example is the spatial diversity technique for optical wireless communications, where the use of intensity modulation/direct detection results in the loss of phase information for spatial gains. Therefore, this book also aims to inspire more interests and further studies in this emerging field.

Author

Dr. Ke Wang is an Australian Research Council (ARC) DECRA Fellow and Senior Lecturer in the School of Engineering, Royal Melbourne Institute of Technology (RMIT University), VIC, Australia. He worked with The University of Melbourne, Australia, and Stanford University, USA, before joining RMIT University. He has published over 110 peer-reviewed papers in top journals and leading international conferences, including over 20 invited papers. He has been awarded several prestigious national and international awards as recognition of research contributions, such as the Victoria Fellowship, the AIPS Young Tall Poppy Science Award and the Marconi Society Paul Baran Young Scholar Award. His major areas of interest include: silicon photonics integration, opto-electronics-integrated devices and circuits, nanophotonics, optical wireless technology for short-range applications, quasi-passive reconfigurable devices and applications and optical interconnects in data centers and high-performance computing.

1

Introduction: Optical Wireless Communications

In the past decade, we have seen a tremendous increase in the demand of high-speed communications, fuelled by a large number of services and applications that require broadband real-time data transmissions. For example, the "retinal" 360° video experience in virtual reality (VR) requires at least 600 Mb/s data transmission and an uncompressed 4K TV with 60 frames/second requires over 10 Gb/s communication speed. These applications and requirements have significantly advanced high-speed communication technologies, especially those providing direct access to end users. For example, optical fiber-based broadband access networks (such as passive optical networks (PONs)) are now widely deployed in many countries. Compared to the wired access approach, the wireless high-speed communication is more attractive since mobility is enabled. Therefore, users are not restricted to one particular location, especially with the availability of high-performance mobile terminals, such as mobile phones and tablets. Such a preference is evidenced by the rapid growth of mobile data traffic, where the data traffic of each mobile increased by about 140% in 2018. It is also predicted that the mobile data traffic will increase by a factor of 5 over the next 6 years [1].

Currently, radio frequency (RF) technologies are widely used to provide wireless access to users, such as the 4G and Wi-Fi technologies. However, such RF technologies are facing a number of limitations, especially the congested RF spectrum, which constrains the achievable communication speed. On other hand, the optical wireless communication (OWC) technology explores the non-congested license-free optical spectrum to provide high-speed wireless data transmissions, and hence, it is a promising solution to meet the users' ever-growing wireless communication speed requirement in the near future. In this chapter, we will briefly overview the OWC systems and technologies for both short- and long-range applications. The rest of this chapter is organized as follows: Section 1.1 provides the basics of OWC; Section 1.2 presents a brief history of OWC communications; the typical long- and short-range OWC applications are summarized in Sections 1.3 and 1.4; and the structure of this book is described in Section 1.5.

1.1 Optical Wireless Communication Technologies

Traditionally, "wireless communication" normally refers to the systems and technologies using the RF, which occupies the electromagnetic spectrum of 30 kHz to 300 GHz. Some widely used frequency ranges include the 1800 and 2100 MHz for 4G mobile communications (in Australia), the 2.4 and 5 GHz for Wi-Fi local area network (LAN) applications, the 3.1 to 10.6 GHz for ultra-wideband (UWB) systems, the 60 GHz millimeter wave range and the 77 GHz for vehicular communications. The lower RF spectrum range nowadays faces the congestion issue, resulting in limited speed, capacity and high interference. The higher RF spectrum range, on the other hand, has high free-space propagation loss, line-of-sight transmission and expensive hardware limitations.

In addition to the RF range, "wireless communication" can also explore other parts of the electromagnetic spectrum, such as the optical frequency range. Normally, the optical spectrum can be divided into the ultraviolet, visible and infrared ranges, and it occupies a broad frequency region from 3 THz (i.e. far-infrared) to over 3000 THz (i.e. near-ultraviolet), as shown in Figure 1.1. An extremely broad bandwidth (practically an unlimited band-width) is available, which indicates that ultrahigh speed wireless communication is achievable. Therefore, the use of this frequency range has great opportunities in the future, especially when considering the emerging high-speed applications and services. The wireless communication systems and technologies exploring this optical spectrum are referred to as OWC systems and technologies.

Compared to traditional RF wireless communications, OWC has a number of unique advantages, such as:

- an extremely broad bandwidth;
- an unregulated bandwidth, which means service providers do not need to pay the costly licensing fee;
- highly secure communications: the optical beams used for transmission are primarily narrow and confined within a certain area, therefore, it is difficult to intercept the communication link;

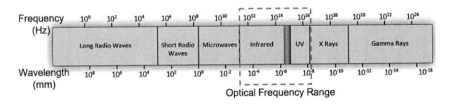

FIGURE 1.1
Electromagnetic spectrum. The optical frequency range is the focus of this book.

- immunity to electromagnetic interference;
- use within RF-hostile environments, such as in aircrafts and hospitals.

In addition to these advantages, typically the intensity modulation (IM) and direction detection (DD) scheme is used in OWC systems. Therefore, OWC does not suffer the multipath fading issue that commonly exists in RF wireless systems. Furthermore, since there is no RF radiation in OWC systems, OWC is widely considered as a non-harmful solution.

The general OWC principle is shown in Figure 1.2. The data to be transmitted first modulates the light source, where the laser diode (LD) or the light-emitting diode (LED) is used. The IM is widely adopted, and it can be realized through both direct modulation or external modulation. The modulated optical signal then passes through the transmitter optics for controlling the beam properties, such as the beam divergence. Then the data-carrying optical beam directly propagates through the wireless link (i.e. free space). At the receiver side, the receiver optics module collects the optical signal beam, and then the photo detector (PD) converts the light signal to the electrical domain as the data output. Typically, the positive-intrinsic-negative (PIN) or avalanche photodiode (APD) is used as the PD, which performs the direct detection. At the receiver side, in addition to the signal light, background light is also collected and detected. The background light normally comes from the sunlight or illumination lamps, and it leads to the additional noise of received data. Each of these function blocks in OWC will be discussed in subsequent chapters.

The OWC systems and technologies can provide high-speed wireless data transmissions over a wide range of distances. Here we roughly divide OWC systems into two categories according to the communication distance – the short- and the long-range types. The short-range OWC refers to the applications where the data transmission distance is up to about one hundred meters, and the long-range type provides communication links from about one hundred meters to over tens of thousand kilometers (e.g. over 300,000 km in NASA Lunar Communications). Some typical short-range OWC systems include the optical interconnects between chips and boards, wireless personal area networks, and underwater communications; and some long-range applications of OWC include terrestrial links, wireless backhauls and space communications. More details about these two types of OWC systems will be provided in Sections 1.3 and 1.4.

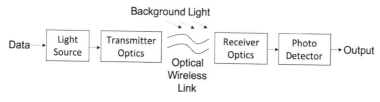

FIGURE 1.2
The general principle of OWC.

With highly collimated beams, accurate transceiver alignments and high-power light sources, the OWC principle has been used in ultra-long range applications, such as Lunar communications and satellite communications, including both satellite-to-satellite and ground-to-satellite scenarios. One example of ultra-long-range OWC application is the Lunar Laser Communication demonstration by NASA [2], where an OWC link provides a transmission speed of 622 Mbps between the Moon and Earth with a distance of over 384,000 km.

1.2 Optical Wireless Communications – Brief History

The use of optical wireless technologies for wireless signal transmissions has a long history, and a brief snapshot of OWC systems is illustrated in Figure 1.3. The early use majorly relies on the signaling through smoke, beacon fires, torches and sunlight. One of the earliest uses of light for communications is along the Great Wall of China, where an ancient OWC system composed of a large number of beacon towers was used for signaling and warning. The number of lanterns and the color of smoke transmits important information, such as the invading of enemy, the size of enemy and the distance of enemy. The relay principle was also used, where beacon towers located with a regular distance collected signal from the previous one and re-transmitted to the subsequent one. Through this way, the message can be successfully transmitted from one end of the Great Wall to the other. Ancient Greeks and Romans also use the similar principle of OWC through fire beacons to transmit key messages around 800 BC. This principle was also used by the American Indians for similar purposes.

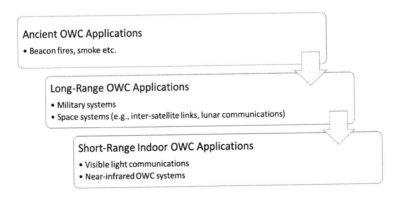

FIGURE 1.3
The brief history of OWC systems and technologies.

In modern times, the experiment conducted by Alexander Bell in 1880 is widely regarded as a milestone in the history of OWC development [3], where voice signals were transmitted via the free space for about 200 m. In the demonstration, at the transmitter side the voice signal caused the vibration of a mirror, and such vibration modulated the sunlight that was projected onto the mirror. Such change of sunlight was detected at the receiver side after wireless transmission, which re-generated the original voice signal.

Although this experiment shows the potential of OWC for wireless signal transmission, due to the device limitations and the intermittent nature of sunlight, the demonstration was highly limited. In addition, due to the advances in the radio communications, the OWC technology did not attract much attention afterwards. However, towards the late twentieth century, the capacity and distance of radio communications encountered the bottleneck and the OWC technology became attractive again. A number of OWC systems were studied and demonstrated in 1960s [4]. For example, researchers at the MIT Lincoln Laboratory demonstrated the transmission of television signal in 1962, where GaAs LED was used as the light source and the signal was transmitted over 48 km. In addition to using LED as the light source, which is an incoherent source, the use of He-Ne laser operated at the wavelength of 632.8 nm was also studied. 190 km transmission of the voice signal was demonstrated in USA in 1963 with He-Ne laser, and later on researchers at the North American Aviation showed the first transmission of television signal using laser over the free space, where up to 5 MHz modulation bandwidth was realized. The interest of using OWC technologies continued in the 1970s, where the first laser OWC link with commercial traffic was built in Japan by Nippon Electric Company (NEC). Still using the He-Ne laser, up to 14 km transmission distance was achieved and the OWC link was full-duplex.

Although a number of technologies were developed and a number of systems were demonstrated in 1960s and 1970s, the OWC results were not satisfactory. This is mainly because of the large divergence of laser beams and the limitations imposed by atmospheric effects. In addition, due to the rapid development of optical fiber communication technologies, especially the availability of low-loss single-mode optical fibers and erbium doped fiber amplifiers (EDFAs), the general interests shifted to optical fiber transmission systems. However, the OWC technology still attracted substantial attention in military and space applications, mainly because of the easy deployment, security, flexibility and large capacity features. At the space application side, the European Space Agency (ESA) and National Aeronautics and Space Administration (NASA) are the major players. For example, NASA investigated the near-Earth OWC systems in the Geosynchronous Lightweight Technology Experiment (GeoLITE) and the Global-Scale Observations of the Limb and Disk (GOLD) programs, and ESA investigated similar principles and systems in the Semiconductor Inter-Satellite Link Experiment (SILEX). Such near-Earth OWC technologies continue to develop in the past decade, and up to 10 Gbps inter-satellite links have been demonstrated [5].

The OWC principle is also popular in deep-space applications. For example, NASA sponsored the Lunar Laser Communication Demonstration (LLCD) project to demonstrate the OWC system with an operation distance of about 400,000 km, which is much longer (more than ten times longer) than the near-Earth systems that have been demonstrated. Up to 622 Mbps OWC from a lunar orbit to a terminal on Earth has already been demonstrated.

In spite of the popularity in military and space applications, the development and application of OWC in the mass market was relatively limited – probably the most successful application of OWC was the IrDA [6], which was a widely used short-range wireless transmission solution. Another success application of OWC in the mass market was serving as the redundant link where fiber optic installations are not feasible. There are main reasons about the lack of OWC popularity: (1) most OWC systems suffer from severe atmospheric effects (i.e. turbulence), leading to system performance degradations such as beam scintillation, and requiring accurate alignment between the transmitter and receiver; (2) the OWC reliability is relatively low, due to both optical path blockages and the impact of heavy fog and (3) current communication systems and technologies already satisfy user requirements mostly, such as the optical fiber communication systems and advanced RF systems (e.g. Wi-Fi and 4G systems).

In the past a few years, thanks to the development of high-performance and low-cost optoelectronic devices, OWC becomes a hot topic again. This trend is further accelerated by the ever-growing bandwidth requirement of end users and the high-speed real-time connectivity required by emerging services. Compared to traditional applications in outdoor and relatively long-range scenarios, the renewed interest in OWC principle focuses more on the short-range and indoor environments. One of the widely studied system in recent years is the indoor OWC system, which can provide direct high-speed wireless connections to users in personal working and living spaces. Both the near-infrared wavelength range and the visible wavelength range have been investigated. More details about the short-range OWC are discussed in Section 1.4.

1.3 Long-Range Outdoor OWC Systems

The OWC principle uses modulated optical signal, which propagates through the free space to the receiver side, to provide data transmissions. In this section, we focus on the application of OWC in long-range systems, with a communication distance ranging from about one hundred meters to tens of thousand kilometers.

One attractive long-range application of the OWC principle is providing the backhauling of 4G and 5G mobile wireless networks, as illustrated in

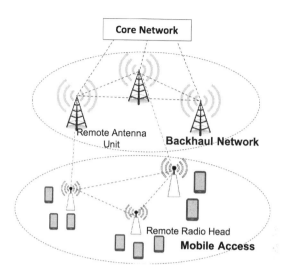

FIGURE 1.4
OWC for mobile backhaul networks.

Figure 1.4. Wireless backhaul networks provide connections between base stations and the core network. In current 4G and future 5G mobile networks, the data rate and capacity supported are significantly higher than previous generations, which puts stringent requirements on backhaul networks. Microwave and optical fiber-based systems have been used to satisfy these needs. However, optical fiber-based backhaul networks require the deployment of fiber, which is costly especially in high-density urbans, and the microwave-based solution faces the limited data rates, licensed spectrum and electromagnetic interference (EMI) issues. To overcome these problems, the OWC principle can be used, where a pair of optical transmitter and receiver directly establishes a wireless link between base stations or between the base station and the core network. Up to 200 m wireless range has been demonstrated, and a reliability of 99.9% is achieved even considering diverse weather conditions such as rainfall, snow and fog [7]. High data rate has also been achieved using the OWC technology, where up to 200 Gb/s has been demonstrated combining the wavelength division multiplexing (WDM) technology [8]. In addition to transmitting baseband signal modulated optical signals, the RF signal for 4G (or 5G) communications has also been transmitted using the OWC technology directly (normally referred as the radio over free space optics or RoFSO technology), eliminating further frequency up-conversion and down-conversion requirements [9].

Due to the easy-deployment advantage (essentially requiring a pair of transmitter and receiver), the OWC technology also finds its application in disaster cases where telecommunication infrastructure is destroyed or unavailable, or the existing communication network collapses [10]. For example, the OWC system was deployed after the 9/11 terrorist attack in New York City to provide data communications and to serve as a redundant link. Because of the easy-deployment advantage, OWC also sees its application in the Wall Street to rapidly provide high-speed communication links for financial corporations that did not already have wired connections [11].

In the outdoor applications mentioned above, the OWC system performance is limited by a number of factors:

- Adverse weather condition: The transmission loss of outdoor OWC systems depends on the weather condition, and additional attenuation can be observed under adverse conditions. This is especially the case with heavy fog, which typically leads to link failure and service interruptions;
- Atmospheric turbulence: Air turbulence widely existing in outdoor environments results in the change of refractive index along the signal propagation path, and it leads to a number of adverse effects, including scintillation, beam broadening and beam wandering. All these effects cause additional noise, higher link loss and instable performance in OWC systems;
- Transceiver alignment: Due to the relatively long link range, narrow and directional light beams are used in outdoor OWC systems. Therefore, precise alignment between transceivers is a necessity. However, precise alignment is challenging in practice, such as the pointing error induced by the building sway when the transceivers are mounted onto buildings. In addition, the beam wandering caused by turbulence also results in the misalignment between transceivers. The transceiver misalignment results in additional loss and service interruptions in the system;
- Ambient light: In outdoor OWC systems, background light from sun exists during the daytime. During the night, illuminations also results in strong background light. Therefore, additional noise is induced in the OWC system;
- Link blocking: As discussed above, narrow signal beam is primarily used. Occasionally, the beam can be partially or fully blocked, such as by birds.

A number of technologies have been studied to overcome the aforementioned limitations in outdoor OWC systems. To reduce the impact of ambient light, researchers have studied the use of far-infrared wavelength band around 10 μm, where the background light spectrum is much weaker [12], and a double-laser differential signaling scheme has also been proposed to improve the OWC system performance when the background noise dominates [13]. To solve the service interruptions and performance degradations caused by misalignment, advanced acquisition, tracking and pointing (ATP) mechanisms have been developed, which point the transmitter in the direction of receiver, acquiring the light signal from the transmitter and tracking the position of the transceivers to maintain the OWC link. Various ATP principles have been proposed and demonstrated, such as gimbal-based, mirror-based, adaptive optics-based and liquid crystal-based technologies [14]. An adaptive optics (AO)-based ATP mechanism is shown in Figure 1.5, where an AO element is used at the receiver side. Due to the impairments during free-space transmission, such as the air turbulence, the signal arriving at the receiver side has distortions. The receiver sensor collects the distortion information and creates a corresponding feedback signal, which is then used to control the AO element to correct the distorted wave [15]. In addition, the multi-input-multi-output (MIMO) principle has also been widely investigated in outdoor OWC systems to overcome the limitations of air turbulence, to provide redundancy to improve system robustness and to improve the system capacity [16–20].

In addition to the applications discussed above, the OWC principle is also widely applied in space communications, including satellite-to-satellite, satellite-to-ground and lunar communications scenarios, where the communication distance is ultra-long [21–24]. One motivation of using light for such applications is that the antenna size can be much smaller than RF systems, due to the short wavelength of optical waves. The directional propagation property of light and the practically unlimited bandwidth available in the optical spectrum are other key reasons of using OWC in space communications. One example of space OWC is the NASA's lunar laser communication demonstration in 2013 [25]. The data rate of 622 Mbps is achieved with an ultra-long transmission distance of 384,600 km to send data between Moon and Earth.

In addition to the deep space application, the OWC technology is also used in the aircraft communications. Airborne terminals can provide continuous data transmissions through tracking, and they have been demonstrated in various types of aircrafts, including both commercial aircrafts and recently, unmanned aerial vehicles (UAV) [26–31]. An application of

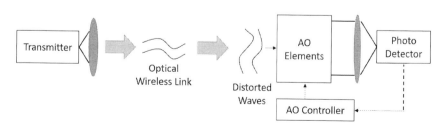

FIGURE 1.5
Adaptive optics-based ATP mechanism.

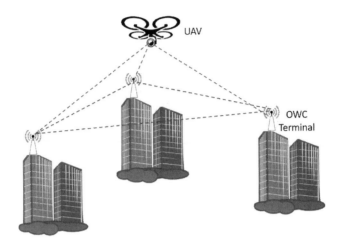

FIGURE 1.6
An OWC system with unmanned aerial vehicle (UAV).

OWC with UAV is shown in Figure 1.6, where the airborne terminal located at the UAV can communicate with the OWC terminals located at the roof of buildings. Such communication links can be used to provide high-speed sensor data transmissions (e.g. real-time video or image transmissions). The UAV can also serve as the relay terminal for the OWC links between buildings to reduce the communication outage possibility, or serve as the service coordinator of the meshed OWC network between buildings.

1.4 Short-Range OWC Systems

The OWC principle has also been applied in short-range applications (with communication distances ranging from centimeters to about one hundred meters) to provide high-speed wireless data transmissions. Actually, a large portion of the renewed interest in OWC technologies recently focuses on short-range applications, mainly including personal and LANs, body area networks, and optical interconnects. This section provides an overview of these rapidly developing fields.

1.4.1 Personal and Local Area Networks

The OWC principle has been proposed and developed to provide high-speed wireless connections in personal and LANs. One such example is shown in Figure 1.7. A broadband access network is deployed and

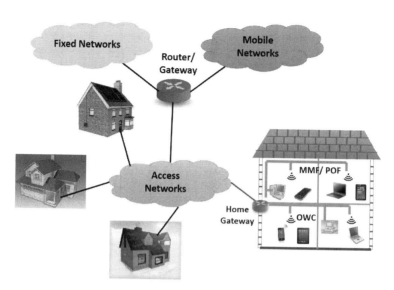

FIGURE 1.7
Application of OWC in personal and LANs.

high-speed connection is delivered to the "doorstep" of users' premises, where an optical line terminal (OLT) provides broadband connectivity to users using the PON access technology. Inside the building, multimode fiber (MMF) or plastic optical fiber (POF) is used to provide high-speed wired connection, due to the easy coupling and the "plug-and-play" advantages. In addition to the wired communication, the OWC system is deployed to cope with the increased demand of wireless connectivity in such indoor environments.

Both the visible and near-infrared wavelength ranges have been studied in OWC-based personal and local area applications. When the visible wavelength is used, the system is referred as the visible light communication (VLC) system [32]. As shown in Figure 1.8, the VLC system uses LED for wireless data transmissions, and hence, it has the key advantage of providing simultaneous illumination and communication functions. This advantage becomes increasingly important with the wide deployment of LEDs for indoor lighting, such as in shopping centers and offices. In addition, due to the spatial confinement of LED lights, cellular type of network (i.e. optical cells) can be used to increase the spectral efficiency whilst keeping the interference at the minimum level [33].

One major limit faced by VLC systems is the relatively small modulation bandwidth of LEDs, partially due to the incoherent nature of lighting. The modulation bandwidth of LEDs is typically in the range of several tens of MHz. To overcome this issue, a number of techniques have been proposed and studied. For example, high order modulation formats with better

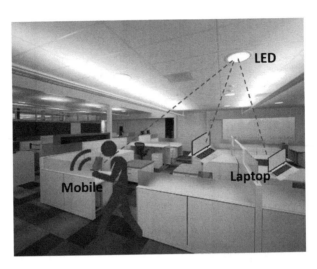

FIGURE 1.8
A VLC system in the indoor office scenario.

spectral efficiency have been investigated, such as the carrierless amplitude and phase (CAP) [34], the orthogonal frequency domain multiplexing
(OFDM) [35] and the discrete multi-tone (DMT) [36] formats. Channel equalization principles have also been applied to further increase the achievable
data rates in VLC systems [37,38]. Another widely used method to increase
the communication data rates in VLC is using WDM, where different colors
of light (i.e. multiple wavelengths) are modulated for parallel data transmissions simultaneously [39]. Combining the high order modulation, channel
equalization and WDM technologies (3 channels with red, green and blue
lights) discussed above, over 4 Gbps VLC has been demonstrated [40], providing high-speed wireless connections in indoor environments to users.

In VLC systems, since multiple LEDs available to provide a satisfying
illumination level, these LEDs can be used for data transmissions simultaneously. Therefore, the MIMO principle can be implemented to further
increase the system data rate and capacity. With a small-scale 2 × 2 MIMO,
over 6 Gbps VLC has been realized [41]. The use of MIMO principle in VLC
systems can also improve the system robustness, thanks to the spatial redundancy provided by multiple signal channels.

In addition to the visible range, the near-infrared wavelength range can
also be used in OWC systems to provide high-speed wireless connections
in personal and LANs. Two wavelength ranges have been widely used – the
850 nm band and the 1550 nm band. The 850 nm band is widely used mainly
due to the availability of low-cost components, such as light sources and
silicon photodiodes, and the compatibility with MMFs. This band is especially popular in the early generation of near-infrared indoor OWC systems.

In recent years, with the maturity of opto-electronic components, the 1550 nm band becomes increasingly popular. This is also because that 1550 nm signals can be distributed with low loss via single mode fibers, and the ambient illumination is much weaker in this band.

In near-infrared indoor OWC systems, lasers are used as light sources. Therefore, compared to VLC systems, one of the key advantages is the much larger modulation bandwidth. Therefore, high-speed wireless connections can be achieved even without advanced modulation format, signal equalization or WDM technologies. For example, with just simple on-off keying (OOK) modulation format, up to 12.5 Gbps communication data rate has been experimentally demonstrated [42].

A typical model of the OWC system in personal environments and LANs is shown in Figure 1.9. An electrical-to-optical (E-O) device (i.e. laser or LED) is modulated by the data to be transmitted, and a data-carrying optical signal is generated. The optical signal then passes through a transmitter optical assembly for beam property adjustments, such as controlling the beam divergence and the propagation direction. The signal beam propagates through the free space. When it arrives at the receiver side, the signal beam is captured by the receiver optical assembly, which typically consists of optical lenses or concentrators for signal collection and focusing. Then the collected optical signal after wireless transmission is converted back to the data in electrical domain by an optical-to-electrical (O-E) device (e.g. photodiode).

In indoor OWC systems, two types of signals are collected and detected by the receiver. The first type is the signal from the direct line-of-sight (LOS) link, which is relatively strong without physical shadowing or blocking. The second type is the signal from multipath links, such as those reflected by walls, the floor and the ceiling. The multipath signal normally

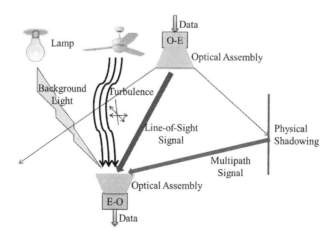

FIGURE 1.9
The general model of indoor OWC systems.

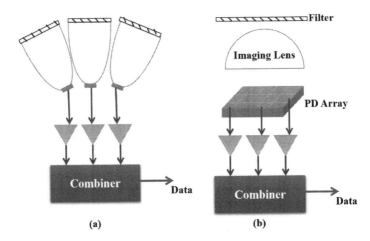

FIGURE 1.10
Advanced receivers for reducing multipath dispersion. (a) The structure of angle diversity receiver and (b) the structure of imaging receiver.

is weaker compared to the LOS signal, and arrives at the receiver with additional delay due to longer propagation distances. Therefore, multipath signals may lead to the multipath dispersion problem, where the transmitted signal arrives at the receiver at different time instances, introducing inter-symbol-interference (ISI) to the system and reducing the effective channel bandwidth. To improve the channel bandwidth due to the multipath dispersion issue, a number of technologies have been developed, especially at the receiver side. For example, the angle-diversity receiver as shown in Figure 1.10a has been studied by using multiple receiving elements that are oriented in different directions [43–45]. The element which collects the signal with the lowest multipath dispersion is then selected and the system bandwidth is improved. Another receiver that can increase the bandwidth in OWC systems is the image receiver shown in Figure 1.10b, where a PD array is used together with an imaging lens. Due to the small size of each PD, multipath signals arriving with different incident angles are filtered out, improving the multipath dispersion [46].

1.4.2 Wireless Body Area Networks and Indoor Localizations

In addition to wireless personal and LANs, the OWC principle is also widely used in the wireless body area network (WBAN), such as the transmission of information collected by bio-sensors inside or on the human body to nearby communication terminals. One key requirement of the WBAN application is the high quality of service, to ensure the correct transmission of health information. RF technologies have been traditionally used in WBAN. However,

RF signals suffer from strong EMI and also cause the EMI issue, and hence, they are prohibited in most medical facilities or hospitals. In addition, the safety of longtime exposure to RF signals is also a concern to users. On the other hand, the OWC technology can solve these limitations whilst providing reliable wireless data transmission in WBAN. For example, an electrocardiography (ECG) sensor is designed in [47], which is incorporated with LED and the sensor data can be transmitted via the OWC system.

Another short-range application of OWC technologies is the indoor user localization, which is highly desired for high-speed data transmissions, indoor navigations, emergency responses and various location-based services, such as the location-based advertisement in shopping centers. The global positioning system (GPS) has been widely used for localization, where a GPS receiver calculates its position by precisely timing the signals sent by GPS satellites high above the Earth [48]. The typical localization accuracy of GPS systems is in the magnitude of several meters and with advanced accuracy enhancement techniques such as carrier phase tracking, the precision can be improved to several tens of centimeters [49]. However, the GPS technology is not suitable for indoor applications, since it suffers from degraded accuracy or even loss of signal due to physical shadowing and multipath dispersion problems [50].

To realize precise indoor localization, several technologies have been developed and the most widely used is the RF-based solution, especially the Wi-Fi (2.4 and 5 GHz bands)-based scheme [51]. In the RF-based indoor localization systems, the location of user is estimated using the received signal strength, the angle-of-arrival and the time-of-arrival information, based on either map-based or non-map-based methods. The typical localization accuracy in indoor spaces ranges from several tens of centimeters to about one meter, and the accuracy is mainly limited by the multipath dispersion of RF signals.

The OWC principle-based indoor localization technology can solve the limited localization accuracy in GPS and RF-based solutions, and both visible and near-infrared wavelengths have been studied. Due to the smaller impact of multipath dispersion, the localization accuracy achieved with OWC technologies is significantly improved. For example, a near-infrared OWC-based indoor localization system using the received signal strength information is demonstrated, and better than 15 cm localization accuracy is achieved with a signal beam size of 1 m [52]. When further using the received signal angle-of-arrival information, the accuracy of better than 5 cm is realized [53]. The height of the receiver is also obtained, providing the three-dimensional (3D) localization function. By tracking the height change of the user and identify abrupt changes, the OWC localization system can be used to detect sudden falls in smart elderly care applications.

In addition to the near-infrared OWC technology, the VLC technology has also been applied in indoor localization applications [54]. Several localization principles have been proposed and demonstrated, mainly including the scene analysis, proximity and triangulation schemes. In the scene analysis method, fingerprint information is collected in various locations in a scene and the user's location is estimated by matching measurements to the pre-collected fingerprints [55]. In the proximity-based VLC localization systems, a dense grid with a large number of reference points is used. Each reference point has a LED for localization signal transmission. When the user receives a signal from the LED located at a reference point, the user is then estimated to be co-located with the reference point [56]. The triangulation principle is also used for indoor VLC localization applications. Typically, multiple LEDs are used and distances from the user to these LEDs are measured. The distance can be measured using different methods, including the received signal strength, the time-of-arrival, the time difference of arrival and the angle-of-arrival-based schemes. Based on the measured distances, the user's location is then estimated [57]. With three LED transmitters, up to 2.4 cm localization accuracy has been demonstrated using the triangulation principle [58].

1.4.3 Optical Interconnects

With the continuous improvement of electronic devices and multi-core architectures, the speed requirement of data communications within data centers and high-performance computing platforms is rapidly increasing, as shown in Figure 1.11. This trend becomes even more obvious with the popularity of data intensive applications. According to the distance that

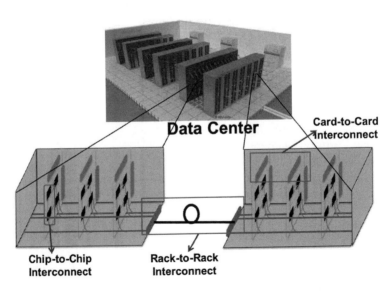

FIGURE 1.11
Interconnects in data centers.

data needs to be transmitted, interconnects can be divided into on-chip, chip-to-chip, card-to-card (also called board-to-board) and rack-to-rack interconnects. Sustained improvement in on-chip and chip-to-chip data interconnects has been demonstrated, such as the silicon photonic interconnects that can achieve over ten times data rate improvement [59,60]. However, the capacity of interconnection between cards and racks has not kept the pace. Conventionally, copper-based interconnects, such as electrical cables, are used for data transmission between cards and racks in data centers and high-performance computing. However, the fundamental limitations, which relate to electric power consumption, heat dissipation, transmission latency, signal loss and electromagnetic interference (or crosstalk), will prevent electrical technologies from scaling up to meet the future demand for high speed and high throughput interconnects.

To overcome the electrical card-to-card interconnect bandwidth limitation, the use of parallel optical short-range links has been proposed and developed. Most of the effort focuses on the use of polymer waveguides [61–63] and MMF ribbons [64,65]. The polymer waveguides-based optical interconnect structure is shown in Figure 1.12a. A VCSEL (i.e. vertical cavity surface-emitting laser) array is driven by a CMOS integrated driver circuit, and the modulated optical beams are then coupled into dedicated polymer waveguides fabricated using the mature and low-cost printed circuit board (PCB) technology. After transmission, the optical signals are coupled out of the waveguides and focused onto a PD array with a lens array to complete data interconnects. Up to 240 Gbps bidirectional parallel optical interconnects have been experimentally demonstrated with a fully integrated low-profile transceiver, which consists of 16-channels VCSEL/PD arrays, lens array and CMOS circuits [63].

The typical structure of MMF ribbon-based optical card-to-card interconnects is shown in Figure 1.12b. An optical transceiver module is

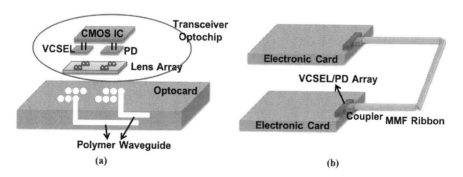

FIGURE 1.12
Structure of optical interconnects. (a) Polymer waveguide-based optical interconnects and (b) MMF ribbon-based optical interconnects.

integrated with the electronic card and it is composed of a VCSEL array and a PD array. Electrical signals to be transmitted from the underlying electronic board first modulate the VCSEL array and then the modulated optical beams are coupled into MMF ribbons and transmitted to the receiver side. Different from conventional telecommunication systems where WDM technology is widely employed to increase the aggregate bit rate, here space-division-multiplexing (SDM) is preferred, which scales up the aggregate bandwidth by using VCSELs at the same wavelength. Multiple fibers in parallel are deployed and each VCSEL interfaces with a dedicated MMF link. This is because this scheme (1) is cost-effective; (2) eliminates the need for complex circuitry for the precise control of the wavelength of the VCSEL elements and (3) increases the aggregate bit rate. Up to 1 Tbps optical interconnects have been developed using 48 parallel channels, and the entire transceiver has been integrated on a single CMOS-compatible holey chip [66].

However, these two kinds of optical card-to-card interconnects can only realize data transmissions between fixed ports. Therefore, such point-to-point interconnection schemes are inherently non-reconfigurable, and their flexibility is very limited. To realize reconfigurable high-speed card-to-card interconnects, the OWC principle has been applied and developed. An example of OWC-based optical interconnect scheme is shown in Figure 1.13. A dedicated optical interconnect module is integrated onto each electronic card (a PCB) to generate data carrying optical beams. Then the optical beams propagate through the free space to the destination electronic card, where the optical signals are collected and detected. Since the data transmission link is the free space, the OWC-based optical interconnects can be dynamically changed to arbitrary directions via a link selection mechanism, such as

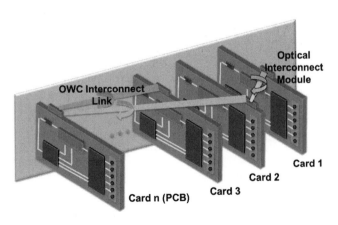

FIGURE 1.13
Structure of an OWC-based board-to-board optical interconnects.

the spatial light modulator, or the MEME integrated steering mirror [67,68]. Therefore, adaptive and reconfigurable optical interconnects with receivers at arbitrary locations within a communication range can be achieved. Based on this principle, up to 120 Gbps reconfigurable OWC interconnects have been demonstrated [69]. Compared to other technologies, the OWC-based interconnect can achieve both dense and reconfigurable optical data communications simultaneously.

1.5 Book Structure

As discussed in the previous sections, the OWC technology has been applied in many different areas, including both long-range outdoor and space applications and short-range indoor and personal area applications. In addition, both the visible and near-infrared wavelength ranges have been studied in OWC systems. Due to the rapid development in recent years, this book will focus on the short-range indoor OWC technologies and systems, especially those using the near-infrared wavelength band. The structure of this book is as follows:

- Chapter 2 provides the general theoretical model of near-infrared indoor OWC systems. It starts with the separate transmitter, receiver and the OWC link models, including both the direct LOS and the diffusive OWC links, and then builds the complete system model. The established model is applied to a high-speed indoor OWC system example. The performance of the system example is analyzed theoretically and then confirmed experimentally. The model of the example system is also used to study the key background light limiting factor in indoor OWCs;

- Chapter 3 introduces the spatial diversity principle in high-speed indoor OWC systems. Both transmitter and receiver diversities are introduced, whilst the focus in on the transmitter diversity scheme. Both repetition coding (RC) and space-time block coding (STBC) transmitter diversity principles are described, which can improve the robustness of high-speed indoor OWC systems against physical shadowing. The RC and STBC schemes are analyzed and compared based on the system model established in Chapter 2. In addition, a delay-tolerant transmitter diversity principle is also introduced to solve the channel delay issue in practical systems;

- Chapter 4 focuses on the WDM technology to increase the data rate and the multi-user access principle to serve multiple users simultaneously in indoor OWC systems. The WDM-based indoor OWC system is studied both theoretically and experimentally.

Both traditional multi-user access techniques, including frequency-division-multi-access (FDMA), code-division-multi-access (CDMA) and time-division-multi-access (TDMA), and the recently proposed time-slot coding (TSA) multi-user access principle are discussed. The robustness of the TSA scheme against imperfect timing is also further studied, and the advanced adaptive loading technique in the TSA scheme to increase the effective system data rate is discussed as well;

- Chapter 5 introduces the photonic integration techniques for near-infrared indoor OWC systems, to solve the bulky size, instable performance and high complexity limitations of current discrete components-based system implementations. The photonic integration technology is briefly reviewed, and we focus on the silicon photonic integration platform due to the compatibility with highly advanced CMOS technology. Commonly used passive and active devices on silicon photonic integrated circuits are reviewed, and the unique silicon integrated beam steering device for indoor OWC systems is discussed in detail. The experimental demonstration of a high-speed indoor OWC system with the silicon photonic integrated circuit is also presented, which demonstrates the capability and great potential of realizing integrated high-speed indoor OWC transceivers and systems;

- Chapter 6 introduces the optical wireless-based indoor localization principle, to provide user tracking and maintain high-speed wireless connectivity in near-infrared indoor OWC systems. The currently available indoor user localization schemes are first reviewed and the major limitations are analyzed. Then the "Search and Scan" principle-based optical wireless indoor localization scheme is described and discussed in detail. The use of received signal strength information to achieve better localization accuracy is also introduced, and the major limiting factor of background light is analyzed theoretically and experimentally. To further provide the height information, the 3D indoor localization principle jointly exploring both signal strength and angle-of-arrival information is also introduced;

- Chapter 7 summarizes this book with the conclusions and discusses another application of the near-infrared optical wireless technology, which is the high-speed optical interconnects with reconfiguration capability in data centers and high-performance computing platforms. Then the chapter concludes the book with discussions of several future needs and possible research directions in the OWC field.

References

1. Available at https://www.ericsson.com/en/mobility-report/future-mobile-data-usage-and-traffic-growth. [Accessed May 2019].
2. Available at http://esc.gsfc.nasa.gov/267/271.html. [Accessed May 2019].
3. F.R. Gfeller and U. Bapst, Wireless in-house data communication via diffuse infrared radiation. *Proceedings of the IEEE*, 1979. **67**: pp. 1474–1486.
4. E. Goodwin, A review of operational laser communication systems. *The Proceedings of IEEE*, 1970. **58**(10): pp. 1746–1752.
5. M. Toyoshima, Trends in laser communications in space, in *GOLCE2010*, Space Japan Review, Tokyo, Japan, 2010.
6. Available at http://www.irda.org/. [Accessed May 2019].
7. V. Jungnickel, D. Schulz, J. Hilt, C. Alexakis, M. Schlosser, L. Grobe, A. Paraskevopoulos, R. Freund, B. Siessegger, and G. Kleinpeter, Optical wireless communication for backhaul and access, in *European Conference on Optical Communications (ECOC)*, IEEE, pp. 1–3, 2015.
8. Z. Zhao, Z. Zhang, J. Tan, Y. Liu, and J. Liu, 200 Gb/s FSO WDM communication system empowered by multiwavelength directly modulated TOSA for 5G wireless networks. *IEEE Photonics Journal*, 2018. **10**(4): p. 7905908.
9. J. Bohata, S. Zvanovec, P. Pesek, T. Korinek, M.M. Abadi, and Z. Ghassemlooy, Experimental verification of long-term evolution radio transmissions over dual-polarization combined fiber and free-space optics optical infrastructures. *Applied Optics*, 2016. **55**(8): pp. 2109–2116.
10. F. Ahdi and S. Subramaniam, Optimal placement of FSO relays for network disaster recovery, in *IEEE ICC*, pp. 3921–3926, 2013.
11. M.A. Khalighi and M. Uysal, Survey on free space optical communication: A communication theory perspective. *IEEE Communications Surveys And Tutorials*, 2014. **16**(8): pp. 2231–2258.
12. T. Plank, E. Leitgeb, P. Pezzei, and Z. Ghassemlooy, Wavelength-selection for high data rate Free Space Optics (FSO) in next generation wireless communications, in *European Conference on Networks and Optical Communications*, IEEE, pp. 1–5, 2012.
13. M.A. Khalighi, F. Xu, Y. Jaafar, and S. Bourennane, Double-laser differential signaling for reducing the effect of background radiation in free-space optical systems. *Journal of Optical Communications and Networking*, 2011. **3**(2): pp. 145–154.
14. Y. Kaymak, R. Rojas-Cessa, J. Feng, N. Ansari, M. Zhou, and T. Zhang, A survey on acquisition, tracking, and pointing mechanisms for mobile free-space optical communications. *IEEE Communications Surveys & Tutorials*, 2018. **20**(2): pp. 1104–1123.
15. W. Thomas and M.A. Vorontsov, *FreeSpace Laser Communications*. New York: Springer, 2004.
16. E. Bayaki, R. Schober, and R.K. Mallik, Performance analysis of MIMO free-space optical systems in gamma-gamma fading. *IEEE Transactions on Communications*, 2009. **57**(11): pp. 3415–3424.

17. I.B. Djordjevic, S. Denic, J. Anguita, B. Vasic, and M.A. Neifeld, LDPC-coded MIMO optical communication over the atmospheric turbulence channel. *Journal of Lightwave Technology*, 2008. **26**(5): pp. 478–487.
18. A. García-Zambrana, C. Castillo-Vázquez, and B. Castillo-Vázquez, Outage performance of MIMO FSO links over strong turbulence and misalignment fading channels. *Optics Express*, 2011. **19**(14): pp. 13480–13496.
19. A.A. Farid and S. Hranilovic, Diversity gain and outage probability for MIMO free-space optical links with misalignment. *IEEE Transactions on Communications*, 2012. **60**(2): pp. 479–487.
20. C. Abou-Rjeily and W. Fawaz, Space-time codes for MIMO ultra-wideband communications and MIMO free-space optical communications with PPM. *IEEE Journal on Selected Areas in Communications*, 2008. **26**(6): pp. 938–947.
21. G.A. Koepf, R.G. Marshalek, and D.L. Begley, Space laser communications: A review of major programs in the United States, *International Journal of Electronics and Communications*, 2002. **56**: pp. 232–242.
22. B. Furch, Z. Sodnik, and H. Lutz, Optical communications in space— A challenge for Europe, *International Journal of Electronics and Communications*, 2002. **56**: pp. 223–231.
23. B. Furch, Z. Sodnik, and H. Lutz, The ESA optical ground station—Ten years since first light, *ESA Bulletin*, 2007. **132**: pp. 34–40.
24. Y. Fujiwara, M. Mokuno, T. Jono, T. Yamawaki, K. Arai, M. Toyoshima, H. Kunimori, Z. Sodnik, A. Bird, and B. Demelenne, Optical inter-orbit communications engineering test satellite (OICETS). *Acta Astronautica*, 2007. **61**: pp. 63–175.
25. Available at http://esc.gsfc.nasa.gov/267/271.html. [Accessed May 2019].
26. V.J. Chan and R.L. Arnold, Results of one GBPS aircraft-to-ground laser-com validation demonstration, in *Proceedings of SPIE 2990, Free-Space Laser Communication*, SPIE, Bellingham, WA, 1997.
27. B. Stotts, Optical communications in atmospheric turbulence, in *Proceedings of SPIE, Free-Space Laser Communications IX 7464*, SPIE, Bellingham, WA, 2009.
28. F. Moll, J. Horwath, A. Shrestha, M. Brechtelsbauer, C. Fuchs, L.A.M. Navajas, A.M.L. Souto, and D.D. Gonzalez, Demonstration of high-rate laser communications from a fast airborne platform. *IEEE Journal on Selected Areas in Communications*, 2015. **33**: pp. 1985–1995.
29. E. Leitgeb, K. Zettl, S.S. Muhammad, N. Schmitt, and W. Rehm, Investigation in free space optical communication links between unmanned aerial vehicles (UAVs), in *International Conference on Transparent Optical Networks*, IEEE, Vol. 3, pp. 152–155, 2007.
30. F. Ahdi and S. Subramaniam, Using unmanned aerial vehicles as relays in wireless balloon networks, in *IEEE International Conference on Communications (ICC)*, IEEE, pp. 3795–3800, 2015.
31. W. Fawaz, C. Abou-Rjeily, and C. Assi, UAV-aided cooperation for FSO communication systems. *IEEE Communications Magazine*, 2018. **56**(1): pp. 70–75.
32. Y. Wang, N. Chi, Y. Wang, L. Tao, and J. Shi, Network architecture of a high-speed visible light communication local area network. *IEEE Photonics Technology Letters*, 2015. **27**(2): pp. 197–200.
33. L. Feng, R.Q. Hu, J. Wang, P. Xu, and Y. Qian, Applying VLC in 5G networks: Architectures and key technologies. *IEEE Network*, 2016. **30**(6): pp. 77–83.

34. P.A. Haigh, A. Burton, K. Werfli, H. Le Minh, E. Bentley, P. Chvojka, W.O. Popoola, I. Papakonstantinou, and S. Zvanovec, A multi-CAP visible-light communications system with 4.85-b/s/Hz spectral efficiency. *IEEE Journal on Selected Areas in Communications*, 2015. **33**(9): pp. 1771–1779.

35. Z. Wang, C. Yu, W.D. Zhong, and J. Chen, Performance improvement by tilting receiver plane in M-QAM OFDM visible light communications. *Optics Express*, 2011. **19**(14): pp. 13418–13427.

36. G. Cossu, W. Ali, R. Corsini, and E. Ciaramella, Gigabit-class optical wireless communication system at indoor distances (1.5–4 m). *Optics Express*, 2015. **23**(12): pp. 15700–15705.

37. L. Zeng, H. Le Minh, D. O'Brien, G. Faulkner, K. Lee, D. Jung, and Y. Oh, Equalisation for high-speed visible light communications using white-LEDs, in *International Symposium on Communication Systems, Networks and Digital Signal Processing*, IEEE, pp. 170–173, 2008.

38. B. Lin, X. Tang, Z. Ghassemlooy, S. Zhang, Y. Li, Y. Wu, and H. Li, Efficient frequency-domain channel equalisation methods for OFDM visible light communications. *IET Communications*, 2017. **11**(1): pp. 25–29.

39. L. Cui, Y. Tang, H. Jia, J. Luo, and B. Gnade, Analysis of the multichannel WDM-VLC communication system. *Journal of Lightwave Technology*, 2016. **34**(24): pp. 5627–5634.

40. M. Zhang, M. Shi, F. Wang, J. Zhao, Y. Zhou, Z. Wang, and N. Chi, 4.05-Gb/s RGB LED-based VLC system utilizing PS-manchester coded nyquist PAM-8 modulation and hybrid time-frequency domain equalization, in *Optical Fiber Communication Conference (OFC)*, Optical Society of America, Los Angeles, CA, pp. W2A–42, 2017.

41. I.C. Lu, C.H. Lai, C.H. Yeh, and J. Chen, 6.36 Gbit/s RGB LED-based WDM MIMO Visible Light Communication System Employing OFDM Modulation, in *Optical Fiber Communication Conference (OFC)*, Optical Society of America, Los Angeles, CA, pp. W2A–39, 2017.

42. K. Wang, A. Nirmalathas, C. Lim, and E. Skafidas, High-speed optical wireless communication system for indoor applications, *IEEE Photonics Technology Letters*, 2011. **23**(8): pp. 519–521.

43. J.B. Carruther and J.M. Kahn, Angle diversity for nondirected wireless infrared communication, *IEEE Transaction on Communications*, 2000. **48**(6): pp. 960–969.

44. G. Yun and M. Kavehrad, Spot diffusing and fly-eye receivers for indoor infrared wireless communications, in *IEEE Conference on Selected Topics in Wireless Communications*, IEEE, pp. 262–265, 1992.

45. C.R. Lomba, R.T. Valadas, and A.M. de Oliveira Duarte, Sectored receivers to combat the multipath dispersion of the indoor optical channel, in *Proceedings of the IEEE International Symposium on Personal, Indoor and Mobile Radio Communications*, Vol. 1, pp. 321–325, 1995.

46. A.P. Tang, J.M. Kahn, and K.P. Ho, Wireless infrared communication links using multi-beam transmitters and imaging receivers, in *IEEE International Conferecen on Communications*, Vol. 1, pp. 180–186, 1996.

47. D.R. Dhatchayeny, A. Sewaiwar, S.V. Tiwari, and Y.H. Chung, Experimental biomedical EEG signal transmission using VLC. *IEEE Sensor Journal*, 2015. **15**(10): pp. 5386–5387.

48. National Research Council (U.S.), *The Global Positioning System: A Shared National Asset: Recommendations for Technical Improvements and Enhancements.* Washington, DC: National Academies Press, 1995.

49. Global Positioning Systems. Available at http://en.wikipedia.org/wiki/Global_ Positioning_System. [Accessed May 2019].

50. F.V. Diggelen, Indoor GPS theory & implementation, in *IEEE Position Location and Navigation Symposium*, New York Institute of Electrical and Electronics Engineers, Piscataway, NJ, 2002.

51. S.A. Golden and S.S. Bateman, Sensor measurements for WiFi location with emphasis on time-of-arrival ranging, *IEEE Transaction on Mobile Computing.* 2007. **6**(10): pp. 1185–1198.

52. K. Wang, A. Nirmalathas, C. Lim, and E. Skafidas, Experimental demonstration of a novel indoor optical wireless localization system for high-speed personal area networks, *Optics Letter*, 2015. **40**(7): pp. 1246–1249.

53. K. Wang, A. Nirmalathas, C. Lim, K. Alameh, and E. Skafidas, Optical wireless-based indoor localization system employing a single-channel imaging receiver. *Journal of Lightwave Technology*, 2016. **34**(4): pp. 1141–1149.

54. A. Naz, H.M. Asif, T. Umer, and B.S. Kim, PDOA based indoor positioning using visible light communication, *IEEE Access*, 2018. **6**: pp. 7557–7564.

55. J. Wang, Z. Kang, and N. Zou, Research on indoor visible light communication system employing white LED lightings, in *IET Communication Technology Applications*, IET, London, UK, 2011.

56. N.U. Hassan, A. Naeem, M.A. Pasha, T. Jadoon, and C. Yuen, Indoor positioning using visible LED lights: A survey. *ACM Computer Survey*, 2015. **48**(2): pp. 20:1–20:32.

57. W. Xu, J. Wang, H. Shen, H. Zhang, and X. You, Indoor positioning for multi-photodiode device using visible-light communications. *IEEE Photonics Journal*, 2016. **8**(1): p. 7900511.

58. H.S. Kim, D.R. Kim, S.H. Yang, Y.H. Son, and S.K. Han, An indoor visible light communication positioning system using a RF carrier allocation technique. *Journal of Lightwave Technology*, 2013. **31**(1): pp. 134–144.

59. Y. Arakawa, T. Nakamura, Y. Urino, and T. Fujita, Silicon photonics for next generation system integration platform, *IEEE Communications Magazine*, 2013. **51**(3): pp. 72–77.

60. H. Subbaraman, X. Xu, A. Hosseini, X. Zhang, Y. Zhang, D. Kwong, and R.T. Chen, Recent advances in silicon-based passive and active optical interconnects. *Optics Express*, 2015. **23**(3): pp. 2487–2511.

61. R. Dangel, U. Bapst, C. Berger, R. Beyeler, L. Dellmann, F. Horst, B. Offrein, and G.-L. Bona, Development of a low-cost low-loss polymer waveguide technology for parallel optical interconnect applications, in *Digest of the LEOS Summer Topical Meetings Biophotonics/Optical Interconnects and VLSI Photonics/WBM Microcavities*, San Diego, CA, 2004.

62. R. Dangel, C. Berger, R. Beyeler, et al., Polymer-waveguide-based board-level optical interconnect technology for datacom applications. *IEEE Transactions on Advanced Packaging*, 2008. **31**(4): pp. 759–767.

63. C.L. Schow, F.E. Doany, C.W. Baks, Y.H. Kwark, D.M. Kuchta, and J.A. Kash, A single-chip CMOS-based parallel optical transceiver capable of 240-Gb/s bidirectional data rates. *Journal of Lightwave Technology*, 2009. **27**(7): pp. 915–929.

64. L.B. Windover, J.N. Simon, S.A. Rosenau, et al., Parallel optical interconnects >100 Gb/s. *Journal of Lightwave Technology*, 2004. **22**(9): pp. 2055–2063.

65. D.M. Kuchta, Y.H. Kwark, C. Schuster, et al., 120-Gb/s VCSEL-based parallel-optical interconnect and custom 120-Gb/s testing station. *Journal of Lightwave Technology*, 2004. **22**(9): pp. 2200–2212.

66. F.E. Doany, Terabit/sec VCSEL-based parallel optical module based on holey CMOS transceiver IC, in *Optical Fiber Communication Conference (OFC)*, OSA, Washington, DC, 2012.

67. K. Wang, A. Nirmalathas, C. Lim, E. Skafidas, and K. Alameh, High-speed free-space based reconfigurable card-to-card optical interconnects with broadcast capability. *Optics Express*, 2013. **21**(13): pp. 15395–15400.

68. M. Aljada, K.E. Alameh, Y.T. Lee, and I.S. Chung, High-speed (2.5 Gbps) reconfigurable inter-chip optical interconnects using opto-VLSI processors. *Optics Express*, 2006. **14**(15): pp. 6823–6836.

69. K. Wang, A. Nirmalathas, C. Lim, E. Skafidas, and K. Alameh, Experimental demonstration of free-space based 120 Gb/s reconfigurable card-to-card optical interconnects. *Optics Letters*, 2014. **39**(19): pp. 5717–5720.

2

Near-Infrared Indoor Optical Wireless Communications: System Architecture, Principle and Modeling

The near-infrared OWC technology has been widely studied to provide high-speed wireless connections in indoor applications, such as in personal working and living spaces (e.g. home, office, shopping centers, airports, etc.) [1,2]. Depending on the OWC link configuration, near-infrared indoor OWC systems can be divided into two major categories: (1) the diffusive link-based and (2) the direct LOS link-based schemes. In this chapter, we will look into the details of such systems, including the system architecture, the general operation principle and the mathematic modeling.

The rest of this chapter is organized as follows: The general near-infrared indoor OWC system architectures, including both diffusive link- and LOS link-based ones, are introduced in Section 2.1; the typical modeling of near-infrared indoor OWC system is given in Section 2.2; an example of high-speed indoor OWC system combining user tracking and limited mobility is discussed in Section 2.3; the key limiting factor background light is analyzed in Section 2.4; and conclusions are given in Section 2.5.

2.1 Near-Infrared Indoor OWC System Architecture

As discussed in Section 1.4.1 and shown by Figure 1.9, an indoor OWC system primarily consists of the light source, the transmitter optics for beam property control and adjustment, the free-space link, the receiver optics for optical signal collection and focusing, and the PD for signal detection. After free-space propagation, two types of optical signals arrive at the receiver side: (1) the signals from direct LOS link and (2) the signals from non-LOS link. An illustration of these two types of signals in an indoor environment is shown in Figure 2.1. The LOS signals directly propagate from the transmitter to the receiver, and hence, it has almost zero signal dispersion and it can carry high-speed data. On the other hand, the non-LOS signals mainly come from the signals reflected by walls, floor, cabinets and ceiling. Since most of

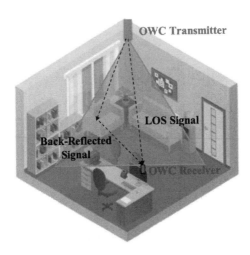

FIGURE 2.1
The LOS signal and back-reflected signal in indoor OWC systems.

the reflections are diffusive, non-LOS signals arrive at the optical receiver from different paths and directions (i.e. various angles of incidence). As a result, optical beams carrying the same data sent by the transmitter may arrive the receiver at different time instances, which lead to the multipath dispersion. Because of multipath dispersion, the non-LOS link generally has limited bandwidth, which restricts the possible high-speed data transmission capability.

In indoor OWC systems, both LOS signals and non-LOS signals can be used for wireless data transmissions. Depending on which type of signal is mainly utilized, indoor OWC systems can be roughly divided into two types: (1) the diffusive link-based and (2) the direct LOS link-based schemes. The general system configuration of both types is shown in Figure 2.2.

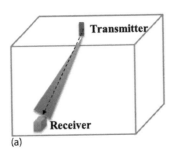
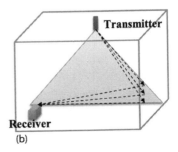

FIGURE 2.2
Typical indoor OWC system configurations: (a) LOS link-based indoor OWC systems and (b) diffusive link-based indoor OWC systems.

2.1.1 Direct LOS Link-Based Indoor OWC Systems

In the LOS link-based near-infrared indoor OWC systems, a relatively narrow optical beam is used to establish a wireless communication link directly between the transmitter and the receiver. Since optical beams naturally reflect back when they strike onto most material surfaces, the LOS link-based OWC system uses several approaches to reduce the collection of back-reflected signals. Two approaches are normally used in current systems: (1) use a narrow light beam for data transmission, and hence, limit possible strikes of signal light onto other surfaces to reduce the existence of back-reflected signals [3] and (2) use a receiver which is only capable of collecting signals within a limited angular range (i.e. limit the field-of-view [FOV] of the receiver), and hence, avoid the collection of reflected signals which arrive from arbitrary incident angles. These two methods are generally combined to some extent to further suppress reflected optical signals.

The LOS link-based near-infrared OWC systems have several advantages:

- *Ultra-broad wireless communication bandwidth*: Since there is very small amount of or even no back-reflected signals collected, there is very limited multipath dispersion in this type of systems. Therefore, the OWC transmission channel has almost unlimited bandwidth.

- *High signal-to-noise ratio (SNR)*: Receivers with relatively limited acceptance angular range is used in LOS-based indoor OWC systems, and hence, only ambient lights that strike onto the receiver within this angular range are collected and contribute to the background light noise. Therefore, the noise in this type system is normally relatively low, and high SNR can be obtained to achieve satisfying performance even at high speeds.

- *High energy efficiency*: Since normally narrow light beams are used or the optical beam only covers a limited area at the receiver side in this type of indoor OWC systems, the total signal power loss is relatively low. Therefore, the transmission power requirement is medium to low, resulting in a high energy efficiency.

- *High communication security*: In LOS-based OWC systems, due to the propagation property of light, the optical signal is mainly restricted within a limited and pre-defined area. Therefore, it is difficult to intercept the wireless communication channel and high security at the physical layer can be provided to users.

- *Low communication latency*: Since the data-carrying light directly propagates from the transmitter to the receiver (i.e. the shortest path) and the light propagates through the free space with high velocity, the transmission delay of LOS-based OWC systems is minimum. Therefore, this type of system is a good option for time-critical applications.

Due to these advantages, the LOS-based near-infrared indoor OWC systems and technologies have been widely studied, and over 10 Gbps data transmissions have been demonstrated [4]. Although it can support very high data rates, the LOS-based indoor OWC systems also have three major limitations: (1) strict alignment between the transmitter and the receiver is needed, due to the use of relatively narrow light beam and the limited reception angular range of the receiver; (2) the mobility that can be provided to users is limited due to the limited signal coverage area; and (3) the wireless communication is vulnerable to physical shadowing, which can be easily caused by moving users that block the LOS link.

2.1.2 Diffusive Link-Based Indoor OWC Systems

In the diffusive link-based near-infrared indoor OWC systems, the optical signals reflected in the indoor environment are utilized to provide the data transmission link between the transmitter and receiver. As shown in Figure 2.3, there are two typical system configurations: (1) a wide optical beam is used to cover the entire communication area, and the reflected signals are collected by a receiver with a large angular reception range; and (2) a transmitter directly pointing at the ceiling is used, and the diffusively reflected signals are collected by the receiver for wireless data communications.

The diffusive link-based near-infrared indoor OWC systems have three major advantages. First, full mobility can be provided to users since the entire communication area is covered by reflected signals, and hence, user can move freely inside the indoor environment without wireless communication service interruptions; second, the strict alignment between the transmitter and the receiver is no longer needed, reducing the system complexity and enabling simple operations; and third, the robustness against physical shadowing and blocking can be achieved, since no LOS signal is required for data transmissions. Due to these unique advantages, the diffusive link-based systems are the most convenient ones for personal and LAN applications.

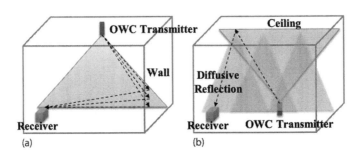

FIGURE 2.3
Two typical configurations of diffusive indoor OWC systems: (a) Diffusive Configuration 1 and (b) Diffusive Configuration 2.

Despite of the advantages discussed earlier, the diffusive link-based indoor OWC system is also mainly limited by four factors:

- *Limited channel bandwidth*: Since the back-reflected signals are used for data transmissions, large multipath dispersion is expected in this type of systems. Therefore, ISI presents at the receiver side, and the available communication bandwidth is limited. The bandwidth in typical indoor office environments is limited to tens of MHz [5].

- *Low SNR*: Since full mobility is provided to users in this type of systems, usually receivers with a large angular reception range are employed to ensure signal collections throughout the communication area. Such receivers also collect ambient light with a large range of incident angles, and hence, strong background light-induced noise is expected, which leads to relatively low SNR.

- *Low energy efficiency*: Since the optical beam needs to cover the entire communication area, the wireless link loss from the transmitter to the receiver is large, generally in the range of 50–70 dB when the coverage area is 5 m [6]. In addition, due to the low-cost consideration, primarily the direct detection is used at the receiver side, which limits the achievable receiver sensitivity. Therefore, a high transmission power is required and the energy efficiency of the system is limited.

- *Safety concern*: As discussed earlier, high transmission power is needed to provide satisfying data transmissions with full coverage. Since lasers are normally used in near-infrared OWC systems, the high transmission power leads to the laser safety concern in the diffusive link-based scheme.

To achieve better communication performance in diffusive link-based OWC systems, a number of techniques have been proposed and studied. To improve the channel bandwidth and reduce the collection of ambient light, a number of advanced receivers have been investigated. One example is the angle-diversity receiver as shown in Figure 2.4a [7]. Here multiple receiving elements are used at the receiver side, and each element consists of an optical filter, an optical concentrator and a PD. The optical filter is employed to only allow the transmission of light within a certain range to reduce ambient light detection, and the optical concentrator is used for efficient signal collection. Typical optical concentrators include the hemisphere lens and the compound parabolic concentrator (CPC), and the use of CPC is illustrated in the figure. Each of the receiving elements has a limited FOV. Therefore, a higher optical gain can be achieved (details will be discussed in Section 2.2), and the multipath dispersion can be reduced since only back-reflected signal within the small FOV is collected. These elements of the

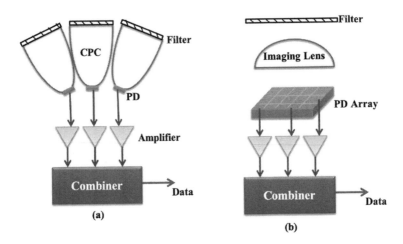

FIGURE 2.4

Advanced receivers for diffusive indoor OWC systems. (a) Angle-diversity receiver and (b) imaging receiver.

angle-diversity receiver are oriented in different non-overlapping directions. And hence, the aggregate FOV of the receiver is still wide to allow full user mobility. After signal filtering, concentration and detection, the converted electrical signals are amplified and fed into a channel combiner. Several combination schemes can be used, such as the select-best principle, where the channel with the highest received signal power or the best SNR is selected for further processing and decision. Based on these principles, the angle-diversity receiver can reduce the background light noise and improve the channel bandwidth simultaneously.

Another advanced receiver proposed for diffusive indoor OWC systems is the imaging receiver shown in Figure 2.4b. The imaging receiver consists of an optical bandpass filter, an imaging optical concentrator (i.e., a lens) and a PD array. The PD array is placed at the focal plane of the lens. The filter rejects most of the out-of-band ambient light, and the imaging lens focuses the signal light onto the PD array. After optical to electrical conversion, a channel combiner is used to select the PD element with the best signal quality. Due to the small size of each PD in the array, only ambient lights with incidence angles similar to the incidence angle of the signal light are collected and detected by the PD element, and hence, the impact of background light noise is significantly suppressed. In addition, since each PD only detects the reflected signals within a small angular range, the multipath dispersion is also reduced. Therefore, similar to the angle-diversity receiver discussed earlier, the imaging receiver is also capable of reducing the background light noise and improving the channel bandwidth simultaneously. The major advantages provided by the imaging receiver are: (1) since only one optical concentrator is used for all detector elements, a more compact

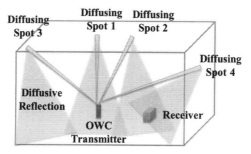

Multi-beam Diffusive Configuration

FIGURE 2.5
The multi-beam transmitter principle for diffusive indoor OWC systems.

system becomes possible; and (2) since a planar PD array is used and this array can be monolithically fabricated using the integrated circuit technology, the cost of receiver can be significantly reduced.

In addition to advanced receivers, a number of advanced transmitter technologies have been proposed and developed as well to improve the data communication capability in the diffusive link-based indoor OWC systems. For example, Yun and Kavehrad proposed a multi-beam transmitter and its structure is shown in Figure 2.5. Here multiple light spots in the ceiling and walls are generated by light sources that point in different directions. Each of the light incident onto the ceiling or walls is reflected diffusively, and hence, multiple diffusing secondary sources are created. Since these diffusing spots are distributed over the ceiling and walls, compared to the case where there is only one diffusing source, the signal power in the multi-beam scheme is more evenly distributed over the communication area and the maximum possible multipath dispersion is reduced. As a result, the system performances such as the SNR and the channel bandwidth can be improved [8]. In addition to using multiple light sources, the multiple diffusing spots can also be generated using the holographic optical diffuser mounted in front of a light source [9], and it reduces the transmitter complexity. The arrangement of multiple diffusing spots has also been studied, and results show that by arranging or generating the diffusing spots in a line (called the line-strip multi-beam transmitter), further system performance improvements can be achieved [10,11].

2.2 Near-Infrared Indoor OWC System Modeling

To design and operate near-infrared indoor OWC systems, it is imperative that the transmitter (i.e. light source), the transmission channel (i.e. free-space channel) and the receiver are well understood. In this section, we will discuss the modeling of indoor OWC systems, including both direct LOS link- and diffusive link-based systems. We will start with the modeling of light sources and detectors, and then model the free-space signal propagation link.

2.2.1 OWC Transmitter

Two types of light sources are widely used as the transmitter in indoor OWC systems – (1) the LED and (2) the laser. Although lasers are more widely used in the near-infrared wavelength range, in this section we will discuss both types to provide a more comprehensive understanding.

The fundamental principle of LED light generation is the spontaneous emission as shown in Figure 2.6a. In the spontaneous emission process, an electron spontaneously returns from an excited state (i.e. in the conduction band of the semiconductor with the energy level denoted as E_c) to a lower energy state (i.e. the valence band of the semiconductor with the energy level denoted as E_v). The energy difference of the electron transition emits a photon with the energy of hv, where $h = 6.626 \times 10^{-34}$ J is the Planck's constant and v is the light frequency. Due to the randomness of photon generation, the photons generated by spontaneous emission are incoherent. In LEDs, electronic excitation is required to achieve sufficient photon generation (i.e. sufficiently strong light emission), and the excitation is generally realized by applying a forward bias to the diode. With the excitation, electrons are excited to the conduction band, which is unstable, and when they return back to the more stable valence band, photons are generated. The wavelength of light (λ) generated is determined by the bandgap E_g between the

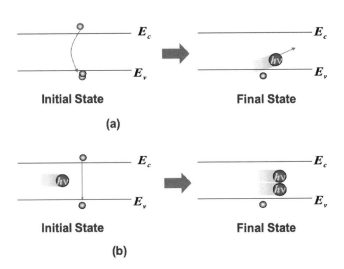

FIGURE 2.6
The principle of photon generation in LED and laser. (a) Spontaneous emission, which is the basis for the LED and (b) stimulated emission, which is the basis for the laser.

conduction band and the valence band (i.e. $E_g = E_c - E_v$), and the wavelength spectra width (defined as the half-power width) is decided by the temperature, which can be expressed as

$$\lambda = \frac{hc}{E_g\,(eV)}\mu m \tag{2.1}$$

$$\Delta\lambda = \frac{1.8\,kT}{hc} \tag{2.2}$$

where c is the light velocity in vacuum and T is the temperature (with the unit of Kelvin). Typical near-infrared LEDs with operation wavelength of 850 nm and 1300 nm have spectral widths of about 60 nm and about 170 nm, respectively.

The typical structure of a LED light source is shown in Figure 2.7, which mainly consists of a forward biased PN junction. Generally, the intensity of light emitted from the LED can be modeled as the generalized Lambertian distribution, where the angular intensity distribution $I(\varphi)$ of the radiation pattern can be modeled using

$$I(\varphi) = \frac{n+1}{2\pi} \times P_t \times \cos^n(\varphi) \tag{2.3}$$

where φ is the angle of light emission with respect to the LED's surface normal, P_t is the total radiated optical power and n is the mode order of Lambertian emission, which describes the shape of the radiated optical beam.

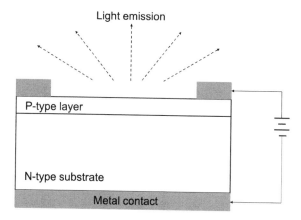

FIGURE 2.7
The basic structure of LED.

The mode order of Lambertian emission is related with the LED radiation semi-angle at half-power $\Phi_{1/2}$:

$$n = -\frac{\ln 2}{\ln(\cos\Phi_{1/2})} \tag{2.4}$$

One limitation of using LED for high-speed indoor OWC systems is the limited modulation bandwidth. The LED modulation bandwidth is mainly limited by three factors: (1) the PN junction capacitance; (2) the injected current (for the excitation of electrons); and (3) the parasitic capacitance. The modulation bandwidth of LED commonly increases with the injected current, and hence, the modulation signal is usually superimposed on a DC bias current to achieve better operation speed. The typical modulation bandwidth of LEDs is in the order of tens of MHz.

In addition to the LED, the laser is also widely used in high-speed infrared indoor OWC systems. Different from LED, which is based on the spontaneous emission of photons, the laser relies on the stimulated emission process. The general principle of stimulated emission is shown in Figure 2.6b. When a photon with the energy $hv = E_g$ is incident onto an excited electron in the conduction band, it triggers the electron to transit to the valence band that has a lower energy level. The energy difference in this transition from the conduction band to the valence band is emitted as the form of a photon. More importantly, this photon generated is identical to the photon that triggers this process, which means the two photons have the same phase and frequency. Therefore, the photons emitted through the stimulated emission process are coherent. This stimulated emission continues, where the generated photon further triggers the emission of more photons in the coherent way, forming the working principle of the laser. Due to the coherence of all photons, the laser generally has a much narrower wavelength spectral width (i.e. laser linewidth) than that of the LED.

The stimulated emission process is an optical gain process, where more photons are generated by the incident photon. However, in typical semiconductors, the photon may lose due to the absorption process, as shown in Figure 2.8. When a photon with an energy $hv \geq E_g$ passes through the

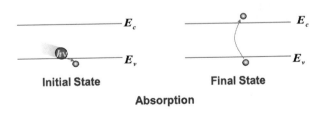

FIGURE 2.8
Absorption of photons in semiconductors.

semiconductor, it can be absorbed by an electron in the valence band, which provides the energy to the electron for jumping to the conduction band. Therefore, the absorption process results in optical loss.

Due to the loss induced by the absorption process, in order to start lasing, the optical gain needs to be at least equal to the optical loss. This is realized through "population inversion", where the number of electrons in the higher energy band is larger than the number of electrons in the lower energy band to create more stimulated emission (i.e. optical gain) than absorption (i.e. optical loss). Population inversion is normally realized through external excitation, which excites electrons into the conduction band. This layer of material is called the active layer.

In addition to population inversion, to produce constant and stable light output, the stimulated emission process needs to be maintained. This is achieved by using the optical feedback – the resonant cavity. The resonant cavity is filled with the medium that can provide optical gain (i.e. with population inversion), and two reflectors are used at both ends of the medium. An example of the resonant cavity is shown in Figure 2.9, where two mirrors are used as reflectors. This type of resonant cavity is known as the Fabry–Perot cavity. Inside the cavity, generated photons via stimulated emission are reflected back by mirrors to stimulate more electrons at the high energy level to jump to the lower energy level, and hence, create more photons to sustain a stable light output.

Compared to the LED, the laser has several advantages. First, coherent light radiation is generated by the laser, and hence, the linewidth is much narrower to reduce the phase noise; second, the laser has much larger modulation bandwidth, ranging from a few hundred MHz to over ten GHz; and third, the optical beam generated by the laser has a smaller divergence, and hence, can be better collimated.

In high-speed near-infrared indoor OWC systems, two types of lasers are widely used as light sources. The first type is the distributed feedback laser (DFB), which is an edge-emitting laser. Different from the Fabry–Perot laser, which uses two mirror reflectors at both ends of the resonant cavity to provide optical feedback, the feedback is provided in a distributed way in the

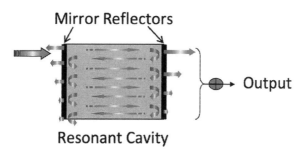

FIGURE 2.9
Laser resonant cavity structure and principle.

DFB laser using distributed Bragg reflectors (DBR). The light reflection of DBR is not from a single point. Instead, light is continuously reflected while it is propagating along the cavity in a round trip. Compared to the single point reflection, the distributed feedback can optimize the optical feedback so that only one single frequency can sustain stimulated emission. Therefore, the DFB laser generally has a single frequency output with narrow linewidth (normally <0.01 nm).

Another widely used laser source in high-speed near-infrared indoor OWC systems is the vertical-cavity surface-emitting laser (VCSEL). In edge-emitting lasers such as Fabry–Perot or DFB lasers, the light propagation direction is co-axis with the active layer. On the other hand, in VCSELs, the light propagates perpendicularly to the active layer, or equivalently, the resonant cavity and the active layer are perpendicular to each other. Due to this structural difference, VCSELs are easier to fabricate in arrays, and hence, they are less expensive than edge-emitting lasers. VCSELs of 850 nm, 1310 nm and 1550 nm are commercially available, and a modulation bandwidth of well beyond GHz is realized. Considering the low-cost advantage, VCSEL is promising for indoor personal and local area applications.

Different from LEDs, the light emitted by lasers are normally modeled as the Gaussian profile, as shown in Figure 2.10a. The electric field distribution of a Gaussian beam on the cross section can be expressed by:

$$E(r) = E_0 exp\left(-\frac{r^2}{\omega_0^2}\right) \tag{2.5}$$

where r is the distance from the center of the beam, E_0 is the electric field at the center of the beam and ω_0 is called the beam waist. ω_0 is defined as the radius at which the amplitude of the electric field drops to E_0/e. Since the

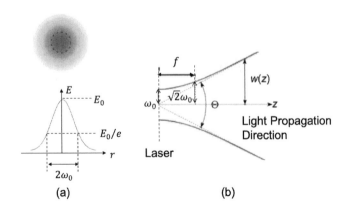

(a)   (b)

FIGURE 2.10
Laser beam electric field distribution. (a) The Gaussian distribution at the cross section and (b) the propagation of laser beam.

intensity of light is the square of electric field, its distribution over the cross section is also Gaussian and can be expressed as:

$$I(r) = I_0 \exp\left(-\frac{2r^2}{\omega_0^2}\right)$$ (2.6)

where $I_0 = E_0^2$ is the intensity of light at the beam center.

When the Gaussian light beam propagates in the free space, which is the z-axis as shown in Figure 2.10b, its field distribution over the cross section at any distance z from the laser is still Gaussian. The Gaussian beam propagation property can be described by three parameters: (1) the beam width $\omega(z)$ at the distance z from the laser (defined as the radius on the cross section at which the amplitude of the electric field drops to E_0/e); (2) the Rayleigh range z_R; and (3) the beam divergence Θ:

$$\omega(z) = \omega_0 \left[1 + \left(\frac{\lambda z}{\pi \omega_0^2}\right)^2\right]^{\frac{1}{2}}$$ (2.7)

$$z_R = f = \frac{\pi \omega_0^2}{\lambda}$$ (2.8)

$$\Theta \approx \frac{\lambda}{\pi \omega_0}$$ (2.9)

where λ is the wavelength of light. The Rayleigh range is the distance from the laser at which the beam width is $\sqrt{2}\omega_0$ (it is also called the f parameter of the Gaussian beam), and the beam divergence characterizes the change of laser beam width whilst propagating.

Due to the limited laser radiation aperture size, the beam directly radiated by laser has a relatively large divergence. Therefore, the beam width becomes larger rapidly with the increase of transmission distance, and the optical power distributed in a unit area drops dramatically. To solve this issue and to provide a controllable optical power distribution after beam propagation, optical lenses are normally employed, as shown in Figure 2.11. When a Gaussian beam radiated by the laser with a beam waist of ω_0 transmits through a lens with a focal length of F, the beam remains Gaussian. If the distance between the laser and the lens is l, then after the lens, we have

$$f' = \frac{F^2 f}{(l - F)^2 + f^2}$$ (2.10)

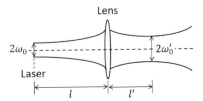

FIGURE 2.11
OWC system transmitter optics for beam divergence control.

$$l' = \frac{l(l-F)+f^2}{(l-F)^2+f^2} \cdot F \tag{2.11}$$

$$\omega_0' = \frac{F}{\sqrt{(l-F)^2+f^2}} \cdot \omega_0 \tag{2.12}$$

where ω_0' is the new Gaussian beam waist after the lens, l' is the distance between the new beam waist and the lens, and f' is the f parameter (i.e. Rayleigh range) of the new Gaussian beam. Beyond the lens, the laser beam remains Gaussian throughout the free-space link. If we assume the beam propagation in the z-direction and set the position of the new Gaussian beam waist as $z = 0$, then the beam width after a free-space distance of z can be obtained by using Eq. (2.7), which is:

$$\omega'(z) = \omega_0' \sqrt{1+\left(\frac{\lambda z}{\pi \omega_0'}\right)^2} \tag{2.13}$$

After obtaining the beam width after free-space propagation, since the beam profile remains Gaussian, both the field and power distributions at the beam cross section can then be calculated using Eqs. (2.5) and (2.6).

2.2.2 OWC Receiver

In indoor OWC systems, after free-space propagation, the data-carrying light signal is normally collected by an optical concentrator and then detected with a PD, as shown in Figure 2.12. An optical bandpass filter is also normally used to reject the out-of-band ambient light. The optical concentrator is normally used to focus the incident light over a large area into a smaller spot (a CPC is shown in the figure), and the PD is used to convert the optical signal into the electrical domain for further processing and decision.

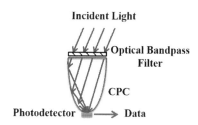

FIGURE 2.12
The typical receiver in indoor OWC systems.

The operation of PD is based on the absorption process shown in Figure 2.8, which occurs in the depletion region of a PN junction. When a photon arrives at the PD, it is absorbed by the electron in the lower energy valence band to jump to the higher energy conduction band if $h\mu \geq E_g$. By applying an external voltage to the PD (i.e. reverse bias the PN junction), the photon-generated electron and hole pair can be swept out to create current, which is called the photocurrent. Since the photocurrent generated is proportional to the photons (which carry the energy of $h\mu$) being absorbed, the PD is a nonlinear square law device that generates an electrical signal (i.e. photocurrent I_p) according to the square of the optical field:

$$I_p = RP_r = R\left(E_r\right)^2 \qquad (2.14)$$

where R is the responsivity of the PD that is defined as the photocurrent generated per unit received optical power, P_r is the received optical power and E_r is the electrical field of the received optical signal.

For high-speed operations, the speed response and the bandwidth of the PD are important, and they are mainly decided by three factors: (1) the transit time of the generated electron and hole pair to be swept through the depletion region; (2) the diffusion time of generated carriers to diffuse out of the PD PN junction; and (3) the RC time constant due to the PD junction capacitance. The RC constant is normally related with the size of the PD. Ordinarily, the drifting of carriers is fast, whilst the diffusion of carriers takes much longer time.

Two types of PDs are widely used in high-speed indoor infrared OWC systems: (1) the PIN type and (2) the APD type. The PIN PD consists of one additional intrinsic region in addition to the normal PN junction, as shown in Figure 2.13a. The intrinsic region is much wider than both P and N regions. Most photons are absorbed in the intrinsic region to create electron and hole pairs. When a reverse bias is applied to the PIN PD, a very large electric field appears across the wide intrinsic region, and hence, the photon generated electrons and

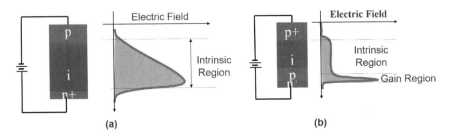

FIGURE 2.13
Structure of two most widely used PDs in indoor OWC systems: (a) PIN detector and (b) APD.

holes can be drifted out with high velocity. Due to the wide intrinsic region, the drifting process dominates and the diffusion time of carriers is significantly reduced. Therefore, the operation speed and bandwidth supported by PIN are usually high, where 10 GHz and higher can be easily achieved.

In addition to PIN, the APD is also widely used. One major difference of APD compared to PIN is that APD has internal carrier gain, through the avalanche process. The basic structure of APD is shown in Figure 2.13b. Compared to PIN, an additional gain region is added. By applying a large reverse bias onto the APD and controlling the doping profile of different regions, a very high electric field is created through the gain region. In APD, the light mostly incidents to the intrinsic region, where photons are absorbed and carriers are generated. When the photon-generated carriers reach the gain region, due to large electric field, they get accelerated and undergo ionizing collision with lattice atoms, which create secondary electron and hole pairs. This process continues to build up more electrons and holes, and hence, provides a gain to the photocurrent. This internal gain process is called the avalanche effect or avalanche multiplication, and the gain can be expressed as the multiplication factor M as:

$$M = \frac{I_{out}}{I_p} \tag{2.15}$$

where I_{out} is the total current at the output of APD and I_p is the primary photocurrent before multiplication. The typical internal gain of APD is in the range of 50–300. Since it takes time for the avalanche multiplication process to build up, normally the speed and bandwidth of APD are lower than that of PIN.

In the process of converting the optical signal to the electrical domain, the PD also adds noise to the output signal. There are two major types of noises caused by the PD. The first type is the dark current noise. Dark current is the current from the PD in the absence of light. Since the photocurrent generated by a signal needs to be at least equal to the dark current to be detected, dark current sets the limit on the minimum detectable signal of a PD. The dark current noise is mainly caused by thermal agitation, and the process of thermally assisted electron and hole pair generation leading to the dark current is a statistical event. Therefore, the dark current noise is a random fluctuation around the mean value of dark current. The mean square dark current noise $\langle i_D^2 \rangle$ can be calculated as:

$$\langle i_D^2 \rangle = 2q \cdot B \cdot I_D \qquad (2.16)$$

where B is the PD bandwidth and I_D is the mean value of the dark current.

The second type of PD noise is the quantum noise, which is also referred as the shot noise. Since the process of absorption leading to the generation of photocurrent is also statistical, there is also a fluctuation around the mean value of photocurrent. This fluctuation is the shot noise, and it can be characterized by the mean square noise current:

$$\langle i_s^2 \rangle = 2q \cdot B \cdot I_p \qquad (2.17)$$

Using Eqs. (2.16) and (2.17), we can obtain the total noise current in PIN:

$$\langle i_{PIN}^2 \rangle = 2q \cdot B \cdot \left(I_p + I_D \right) \qquad (2.18)$$

In addition to the dart current noise and the shot noise, there is an additional noise source in APD, which is the avalanche noise. The avalanche noise is caused by the fact that the process of avalanche multiplication is also a random process, and it multiplies the total of the dark current noise and the shot noise by the an excess noise factor x. Therefore, the mean square noise current of the APD can be calculated by:

$$\langle i_{APD}^2 \rangle = 2q \cdot B \cdot \left(I_p + I_D \right) \cdot M^{2+x} \qquad (2.19)$$

In addition to the PD, the optical concentrator is also widely used in the receiver of indoor OWC systems. The optical concentrator is capable of focusing the light incident onto its input surface to the output surface. Since the input surface is larger than the output, optical concentration gain is provided. Two types of optical concentrators are used: (1) the non-imaging and (2) the imaging types; and the non-imaging optical concentrator is more widely used. The optical gain of an ideal non-imaging concentrator is:

$$g_c(\theta) = \begin{cases} \dfrac{n^2}{\sin^2\theta_{FOV}}, & 0 \le \theta \le \theta_{FOV} \\ 0, & \theta > \theta_{FOV} \end{cases} \qquad (2.20)$$

where θ is the incident angle, n is the internal refractive index of the optical concentrator and θ_{FOV} is the FOV of the concentrator, which characterizes the light angular collection capability. It can be seen from Eq. (2.20) that the optical gain of the concentrator increases when the FOV is reduced. However, in indoor OWC systems, due to the possible movement of users, a large receiver FOV is desired to ensure the collection of light signal even at room corners. To solve this issue, the angle-diversity receiver briefly discussed in Section 2.1.2 and shown in Figure 2.4a is proposed. By using multiple optical concentrators with relatively narrow FOV, high optical gain and a large aggregate FOV can be achieved simultaneously.

2.2.3 OWC Free-Space Channel

In indoor infrared OWC systems, the signal transmission link is the free-space. As shown in Figure 2.1, two types of OWC channel exist in typical indoor environments, namely the LOS channel and the back-reflected channel, which is also normally referred as the non-LOS channel. In this section, we will establish the models for both types of channels.

To characterize the optical wireless channel, the model shown in Figure 2.14 is used. Here $x(t)$ is the transmitted optical signal modulated by the data, $h(t)$ is the channel impulse response, $n(t)$ is the added channel noise and $y(t)$ is the signal arriving at the optical receiver side, which will be filtered by an optical bandpass filter, focused by an optical concentrator, and then converted to the electrical domain by a PD. The major channel noise source is the background noise due to ambient lights.

Signal propagating via the non-LOS link in OWC systems is subject to multipath propagation. Similar to RF systems, the multipath dispersion causes the electrical field of the light signal to suffer from amplitude fades, and the amplitude fades have the scale of a wavelength. Fortunately, OWC receivers generally have a surface area that is much larger than the wavelength scale. For example, for near-infrared indoor OWC systems, an optical receiver with a surface area of 1 mm² is about one million times of the wavelength squared. Therefore, the small-scale electric field amplitude fades caused by the multipath propagation in non-LOS link are averaged out at the optical receiver, and the OWC system does not experience noticeable multipath fading problem.

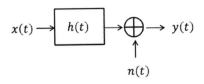

FIGURE 2.14
The channel model of indoor OWC systems.

In indoor near-infrared OWC systems, the data transmission rate is high and the user movement speed is typically limited to walking speed (about 4–5 km/h). Therefore, the communication channel can be considered as quasi-static. Using the channel model shown in Figure 2.14, the optical signal arriving at the receiver side after free-space transmission can be written as:

$$y(t) = x(t) \otimes h(t) + n(t) = \int_{-\infty}^{\infty} x(\tau) h(t - \tau) d\tau + n(t) \qquad (2.21)$$

where the transmitted signal $x(t)$ is the optical signal and it is represented by the power rather than the amplitude. Hence, the channel impulse response $h(t)$ is the power response as well.

In a LOS link, since there is no multipath propagation, all signals arrived at the receiver at the time t are transmitted at the time $t - d/c$, where d is the free-space distance between the transmitter and the receiver, and c is the speed of light. In this case, the channel impulse response essentially represents the total free-space link power loss, and the received LOS signal can be described as:

$$y(t) = x\left(t - \frac{d}{c}\right) h(t) + n(t) \qquad (2.22)$$

For non-LOS links, signals arriving at the receiver experience back-reflections. The ray-tracing algorithm can be used to obtain the channel impulse response, taking both the time delay and the link power loss into consideration [12]. This algorithm traces every back-reflection independently and then sums all together to obtain the complete impulse response. Therefore, multiple back-reflections can be included in the model. Assume a single signal transmitter Tx (considered as a point source) and receiver Rx in a room, the channel impulse response using the ray-tracking algorithm can then be written as:

$$h_{non\text{-}LOS}(t, Tx, Rx) = \sum_{i=0}^{\infty} h_{non\text{-}LOS}^{i}(t, Tx, Rx) \qquad (2.23)$$

where $h_{non\text{-}LOS}^{i}(t, Tx, Rx)$ is the impulse response due to the ith back-reflection ($i = 0$ refers to the signal propagation from the source to the 1st reflecting surface). In indoor OWC systems, the reflecting surface can be considered as a secondary source. The radiation pattern of these secondary sources can be modeled as a generalized Lambertian source, which is described by Eq. (2.3).

The ray-tracing algorithm can be applied to multiple signal transmitters. With T transmitters, the channel impulse response can be described as:

$$h_{non\text{-}LOS}(t, Rx) = \sum_{j=1}^{T} \sum_{i=0}^{\infty} h_{non\text{-}LOS}^{i}(t, Tx_j, Rx)$$ (2.24)

where Tx_j is the jth signal transmitter.

Comparing the signal propagating through the LOS link as described by Eq. (2.22) and the non-LOS link as described by Eq. (2.24), signals from the non-LOS link experience significant multipath propagations. Therefore, signals from various non-LOS links with different reflections reach the receiver at different time instances, leading to the signal dispersion. The signal dispersion results in ISI, and hence, limits the maximum channel bandwidth. Because of this, the LOS link is normally preferred when high-speed communications are required.

In both LOS and non-LOS channel-based indoor near-infrared OWC systems, the major additive noise $n(t)$ as shown in Figure 2.14 from the free-space link is the background noise due to ambient lights. The major contributors to background light include sunlight, incandescent lamps and fluorescent lamps. Due to the popularity of LEDs, they also become a major background light source in OWC systems. The background light results in additional photocurrent at the optical receiver, and hence, generates shot noise after the PD. The shot noise (mean square noise current) due to background light can be calculated as:

$$\left\langle i_s^2 \right\rangle = 2q \cdot B \cdot R \cdot P_{bn}$$ (2.25)

where R is the PD responsivity and P_{bn} is the received background light power. The received background light includes that comes from sources directly and that comes from back-reflections. For the background light from sources directly, it can be calculated using the LOS channel model by considering the power loss; and for the background light that comes from back-reflections, the ray-tracing method described by Eq. (2.24) can be used.

Since the noise is proportional to the received background light power, it can be suppressed by reducing the background light collected at the receiver. This can be realized by using an optical bandpass filter. However, since the passband (including both the central wavelength and the bandwidth) of the filter depends on the signal incident angle, in indoor OWC systems with moving users, the filter bandwidth needs to be selected at tens of nanometers.

2.3 High-Speed Near-Infrared Indoor OWC System Example

In the previous section, the three major parts of an indoor OWC system, which are the transmitter, the receiver and the free-space link, have been discussed and modeled. In this section, we apply those principles and models in a high-speed near-infrared indoor OWC system example, to build the complete system model and investigate the system performance. This section is organized as follows: Section 2.3.1 provides the system description; Section 2.3.2 presents the system operation principle; Section 2.3.3 establishes the theoretical system model; and we analyze the system performance in terms of bit-error-rate (BER) and SNR in Section 2.3.4. This system is also experimentally demonstrated, and the experiments will be discussed in Section 2.3.5.

2.3.1 High-Speed Near-Infrared Indoor OWC System Structure

The high-speed near-infrared indoor OWC system example considered in this section is a practical office environment as shown in Figure 2.15. The room size is $10 \times 8 \times 3$ m. The optical receiver of system (i.e. end users), either fixed or mobile, is supposed to be placed on a plane that is 1 m above the ground (i.e. $z = 1$), and this plane is called the communication floor (CF). This room consists of eight rectangular cubicles with surfaces parallel to the room walls. The size of each cubicle is $2.5 \times 3 \times 1.5$ m. All the partitions are opaque so the signals incident on them are either absorbed or blocked. The cubicles are also equipped with tables and chairs. For illumination purposes, the room is equipped with eight 100 W tungsten floodlights (~550 lux illumination). These lamps are positioned at coordinates of $(x, y, z) = (2, 2, 3)$, $(4, 2, 3)$, $(6, 2, 3)$, $(8, 2, 3)$, $(2, 5, 3)$, $(4, 5, 3)$, $(6, 5, 3)$ and $(8, 5, 3)$. It is obvious that in such scenario, there is strong background light.

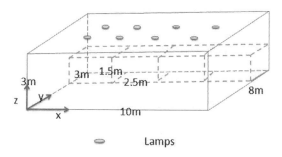

FIGURE 2.15
The practical office environment considered for the OWC system modeling.

In this system, near-infrared lasers are used as light sources since they have broader modulation bandwidth. Optical lenses are also employed at the transmitter side, to adjust the beam divergence and hence, control the signal beam width at the user side. At the receiver, the CPC is used as the optical concentrator for efficient signal collection. An optical bandpass filter with 30 nm passband $(B_{filter} = 30\,nm)$ is used in front of the CPC to suppress out-of-band background light. The relatively broad passband is selected to allow user movement since the filter response varies with the optical signal incident angle. The PIN PD is used after the CPC for O-E signal conversion.

To achieve over Gbps wireless communication, the direct LOS link is selected for data transmission. The maximum radiation power of laser is limited by the laser eye and skin safety regulations. Therefore, considering the link loss and the requirement of providing mobility to users throughout the room, multiple optical transmitters are employed here. The use of multiple transmitters can also help to avoid possible LOS signal shadowing due to physical obstacles (e.g. cubicle partitions) by appropriate selection of the transmitter locations. To reduce the number of transmitters needed to provide signal coverage over the entire room, instead of using the static transmitter that radiates signal towards a fixed direction, the beam steering method has been proposed [13]. The detailed operation principle of beam steering-based indoor OWC system will be discussed in the next section, and the general idea is providing limited signal coverage according to user's location. In the office environment considered here, four optical transmitters with beam steering capability are used, and they are located at $(x_t, y_t, z_t) = $ (2.5, 3, 3), (7.5, 3, 3), (2.5, 5, 3) and (7.5, 5, 3), respectively.

2.3.2 High-Speed Near-Infrared Indoor OWC System Operation Principle

As mentioned in the previous section, the beam steering-based transmitters are used in the high-speed near-infrared indoor OWC system example, which can reduce the number of transmitters needed whilst providing LOS link over the entire room. The structure of optical transmitter with steering mirror is shown in Figure 2.16. The steering mirror is used after the lens to change the optical beam propagation direction, and the micro-electro-mechanical-system (MEMS) technology-based mirror is selected here since it is simple in tuning mechanism (i.e. signal reflection) and has the potential to be mass fabricated. Since the MEMS-based steering mirror is close to the laser and the lens, it can be of small size and light weight. Therefore, a wide tuning range and a reasonable tuning speed (typically in the *ms* scale) can be achieved.

Due to the laser safety regulation, the maximum optical transmission power in the system is limited, and normally it is only practical to cover a limited indoor area. Therefore, here the combination of limited coverage

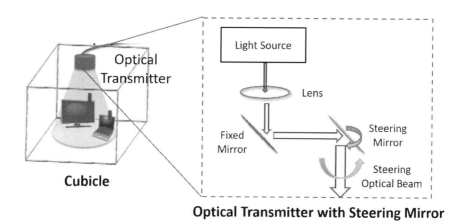

Optical Transmitter with Steering Mirror

FIGURE 2.16
The structure of steering mirror-based optical transmitter.

and beam steering function is used to provide user mobility. This is enabled by the indoor localization technology, which provides the location information of the user. According to the location of the user, an optical signal beam with moderate divergence is created to cover the user's location and its immediate surroundings.

The general operation principle of the high-speed indoor OWC system example with steering mirror is as follows:

1. User localization, which provides the location information of the user. This function can be achieved through a number of technologies, such as the GPS-based, WiFi-based, and optical wireless-based solutions. The user localization function will be discussed in Section 6.

2. Link establishment, where data-carrying optical signal beam is guided by the steering mirror to cover the location of the user and the surrounding area, to establish high-speed LOS link and to start OWC.

3. Data communication: When the user is moving inside the area covered by signal beam, LOS link is always available for high-speed OWC.

4. User tracking and link reconfiguration: The localization function tracks the movement of the user, and when the user moves out of the originally optical signal coverage area, the steering mirror inside the optical transmitter guides the signal beam to the new location to reconfigure the OWC link and to maintain high-speed LOS wireless connectivity.

Through these steps, high-speed OWC can be provided to the user over the entire room.

2.3.3 High-Speed Near-Infrared Indoor OWC System Model

Based on the high-speed indoor OWC system structure and the operation principle described in the previous two sections, here we apply the basic modeling principles discussed in Section 2.2 to establish the theoretical analysis model of the system. The detailed process is as follows:

1. Optical transmitter: Since laser is used in the system example, the optical signal beam radiated can be modeled as Gaussian. The beam divergence and the resulting beam width can then be controlled by selecting the distance between the light source and the lens (l in Figure 2.11) according to Eqs. (2.7)–(2.13). The light is then guided by the steering mirror to the desired direction. For simplicity, here we assume that the half beam divergence is θ_t ($\theta_t = 1/2\Theta$) and the loss due to beam steering is negligible. The negligible loss assumption is usually valid since MEMS mirrors can provide high light reflection coefficient through appropriate coating.

2. LOS link propagation: Assume the user is located at $(x, y, z) = (x_u, y_u, z_u)$ and the optical transmitter is located at $(x, y, z) = (x_t, y_t, z_t)$. The LOS signal propagation distance is:

$$D_{LOS} = \sqrt[2]{(x_u - x_t)^2 + (y_u - y_t)^2 + (z_u - z_t)^2} \qquad (2.26)$$

Assume the signal propagation distance is much larger than focal length of the lens used in the transmitter, then the beam width at the receiver side can be approximated as:

$$\omega_u \approx D_{LOS} \cdot \tan\theta_t + \omega_0 \approx D_{LOS} \cdot \tan\theta_t \qquad (2.27)$$

where ω_0 is the beam waist at the transmitter. If the optical receiver points up vertically, then the signal incident angle is:

$$\theta_u = \arctan\frac{\sqrt{(x_u - x_t)^2 + (y_u - y_t)^2}}{|z_u - z_t|} \qquad (2.28)$$

Due to the limited user localization accuracy, the signal beam center can be different from the user's location, and the center is assumed to be (x_c, y_c, z_c), where $z_c = z_u$. The distance between the signal beam center and the user's location is then:

$$D_{uc} = \sqrt{(x_u - x_c)^2 + (y_u - y_c)^2} \qquad (2.29)$$

Since the beam profile at the user side is still Gaussian, the signal intensity at the receiver can be calculated as:

$$I_u = I_{u0} exp\left(-\frac{2D_{uc}^2}{\omega_u^2}\right) \tag{2.30}$$

where I_{u0} is the beam intensity at the beam center after free-space propagation, and it is related with the original beam center intensity at the transmitter side (i.e. I_0) by:

$$I_{u0} = I_0\left(\frac{\omega_0}{\omega_u}\right)^2 \tag{2.31}$$

The peak intensity at the transmitter side is related with the total transmission optical power P_t by:

$$I_0 = \frac{2P_t}{\pi\omega_0^2} \tag{2.32}$$

3. Optical receiver: Assume the optical bandpass filter has an in-band transmission efficiency of T_f and the surface area of the PD is S_{PD}. If the FOV and the refractive index of the optical concentrator used in front of the PD are θ_{FOV} and n, respectively, using Eq. (2.20), the input surface area of the concentrator is:

$$S_c = \frac{n^2}{\sin^2\theta_{FOV}} \cdot S_{PD} \tag{2.33}$$

Since limited mobility is needed in the system to allow user movements within a certain region, it is practical to assume that the beam width at the user side is much larger than both the input surface area of optical concentrator and the surface area of the PD (i.e. $\omega_u \gg S_c$). Therefore, using Eqs. (2.26) through (2.33), the signal power collected by the receiver can be expressed as:

$$P_r = \begin{cases} \frac{2P_t}{\pi\omega_u^2} \cdot exp\left(-\frac{2D_{uc}^2}{\omega_u^2}\right) \cdot T_f \cdot \frac{n^2}{\sin^2\theta_{FOV}} \cdot S_{PD}, & 0 \le \theta_u \le \theta_{FOV} \\ 0 & , \theta_u > \theta_{FOV} \end{cases} \tag{2.34}$$

In this model, the free-space propagation attenuation is ignored, since the transmission distance in indoor environments is limited to only a few meters to tens of meters. Using Eq. (2.34), the received signal power can be calculated.

After filtering and concentration, the optical signal is then converted to the electrical domain using the PIN PD. The pre-amplifier is commonly used directly after PIN to convert the photocurrent to a voltage signal, which is easier to process. Therefore, the dominant noises here are the shot noise induced by the ambient light and the receiver pre-amplifier-induced noise [14]. Assume here we use the simplest on-off-keying (OOK) format for data modulation, then the noise variance σ_0^2 and σ_1^2 associated with the transmitted signal "0" and "1", respectively, can be calculated by

$$\sigma_0^2 = \sigma_{pr}^2 + \sigma_{bn}^2 + \sigma_{s0}^2 \tag{2.35}$$

$$\sigma_1^2 = \sigma_{pr}^2 + \sigma_{bn}^2 + \sigma_{s1}^2 \tag{2.36}$$

where σ_{pr}^2 represents the preamplifier noise variance component, σ_{bn}^2 represents the background light-induced shot noise variance, and σ_{s0}^2 and σ_{s1}^2 represent the shot noise variance components associated with signal "0" and "1", respectively. Normally the signal-dependent noise is much smaller than the other two noises, so it will be ignored in the following analysis and simulation. Hence, the noise variance in the system can be simplified to:

$$\sigma_0^2 = \sigma_1^2 = \sigma^2 = \sigma_{pr}^2 + \sigma_{bn}^2 \tag{2.37}$$

The preamplifier is normally the field-effect transistor (FET)-based transimpedance amplifier (TIA). Major noise sources are the Johnson noise due to FET channel conductance, the Johnson noise from the load or feedback resistor, the shot noise arising from gate leakage current and the *1/f* noise. The preamplifier noise variance generally can be written as [15]:

$$\sigma_{pr}^2 = \left(\frac{4kT}{R_F} + 2qI_L \right) I_2 B + \frac{4kT\Gamma}{g_m}(2\pi C_T)^2 A_F f_c B^2 + \frac{4kT\Gamma}{g_m}(2\pi C_T)^2 I_3 B^3 \tag{2.38}$$

where B is the system bit rate, A_F is the weighting function (A_F = 0.184 for the non-return-to-zero (NRZ) data format and A_F = 0.0984 for the return-to-zero (RZ) format with 50% duty cycle), I_L is the total leakage current (i.e. the sum of FET gate current and PD dark current), g_m is the FET transconductance, Γ is a noise factor associated with channel thermal noise and gate-induced noise in the FET, C_T is the total input capacitance that consists of both PD and stray capacitance, f_c is the *1/f* corner frequency of the FET, I_2 and I_3 are the weighting functions that are dependent only on the received optical pulse shape and the equalized output pulse shape, R_F is the feedback resistance, k is the Boltzmann's constant, T is the absolute temperature in the unit of kelvin, and q is the electron charge. In most cases, the FET gate leakage and the *1/f* noise can be neglected [16], and hence, the preamplifier-induced noise variance can be simplified to:

$$\sigma_{pr}^2 = \frac{4kT}{R_F} I_2 B + \frac{4kT\Gamma}{g_m}(2\pi C_T)^2 I_3 B^3 \qquad (2.39)$$

The background light-induced shot noise variance can be calculated by:

$$\sigma_{bn}^2 = 2q \cdot R \cdot P_{bn} \cdot I_2 \cdot B_f \qquad (2.40)$$

The background light power collected at the receiver side originates from the eight lamps within the room, as shown in Figure 2.15. Each of them can be modeled as a generalized Lambertian source with a radiant intensity described by Eq. (2.3). To calculate the received background light power, the ray-tracing principle described by Eq. (2.24) is used, where the contributions from lamps are considered independently. For the back-reflection modeling, the ceiling, floor and walls are divided into small reflecting elements. Each element is first considered as a receiver that collects background light, and then it is considered as a secondary point source that re-emits into the free space. The re-emitted power equals to the power collected multiplied by the reflectivity ρ_r, and the re-emitted light pattern is generalized Lambertian. Here, we consider up to three back-reflections, which is sufficiently accurate especially when considering the use of optical bandpass filter at the receiver side. For the first reflection, the size of each reflecting element is selected at 5×5 cm; and for the second and the third reflections, each element is selected at 20×20 cm. The Lambertian pattern order is $n = 1$, and the reflectivity of the wall, floor and ceiling is 0.8, 0.3 and 0.8, respectively [17].

In many practical indoor office environments, florescent lamps are widely used instead of the tungsten floodlights as assumed in Section 2.3.1. This kind of lamps can also be modeled as a Lambertian source [17]. However, the mode number associated is $n = 31$. Due to the high mode number, the background light is more evenly distributed over the entire room. Therefore, at locations directly under lamps, smaller background power is collected by the receiver compared to the case considered here, where tungsten floodlights create large background power at these locations. As a result, the impact of background light from fluorescent lamps is not so obvious at these locations, and we only consider tungsten floodlights for the worst-case scenario.

Combining the optical transmitter, LOS link and optical receiver models discussed earlier, the complete model of the high-speed indoor OWC system example with steering mirror can be established, and the system performance can then be analyzed either theoretically or through simulations.

2.3.4 High-Speed Near-Infrared Indoor OWC System Simulation and Analysis

In this section, we will simulate the indoor OWC system example using the model established in the previous section. With the room configuration as described in Section 2.3.1 and when the receiver is placed at the CF with $z = 1$, the simulated received background light power and the noise variance over the entire room are shown in Figure 2.17.

It can be seen from Figure 2.17a that the maximum received background light power occurs at the locations of (2, 2, 1), (4, 2, 1), (6, 2, 1), (8, 2, 1), (2, 5, 1), (4, 5, 1), (6, 5, 1) and (8, 5, 1). This is because these positions are directly under a lamp. In other positions, the received background light power is much smaller (<1.5 μW (-28.2 dBm)). The noise variance distribution shown by Figure 2.17b follows similar trend, where the peak of the noise variance also occurs directly under lamps.

To analyze the system communication performance, the SNR is an important parameter. In the system example where the OOK modulation is used, the SNR can be calculated as:

$$SNR = \left(\frac{R \times (P_{s1} - P_{s0})}{\sigma_0 + \sigma_1} \right)^2 \tag{2.41}$$

where P_{s0} and P_{s1} are the received powers of signal "0" and "1", respectively. Therefore, $P_{s1} - P_{s0}$ represents the eye opening of the received optical signal and $1/2(P_{s1} + P_{s0})$ represents the received signal power. Here we assume the NRZ data format is utilized, and hence, the BER of the system can then be expressed by:

$$BER = \frac{1}{2} erfc\left(\frac{SNR}{2\sqrt{2}} \right) \tag{2.42}$$

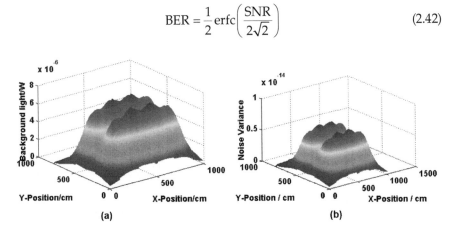

(a)   (b)

FIGURE 2.17
Simulation results of: (a) the received background light power and (b) the noise variance in the system.

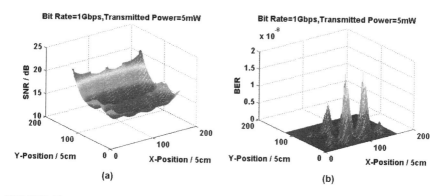

FIGURE 2.18
1 Gbps indoor OWC system simulation results. (a) SNR and (b) BER.

Based on Eqs. (2.41) and (2.42), here we simulate the system SNR and BER performance. When the transmitted power is 5 mW (7 dBm), which is within the laser safety limit, the bit rate is 1 Gbps and the signal beam covers 2 m area surrounding the user for limited mobility; the SNR and BER performances of the system over the entire room are shown in Figure 2.18. It can be seen that at (2, 2, 1), (4, 2, 1), (6, 2, 1), (8, 2, 1), (2, 5, 1), (4, 5, 1), (6, 5, 1) and (8, 5, 1), the SNR is lower and the BER is higher than the surrounding area. This is due to the stronger background light as shown in Figure 2.17a. Fortunately, the SNR difference over the entire room is not large (within 6 dB) since the optical transmitters are also located near these locations. It is also clear from the figure that the system performance is asymmetrical, resulting from the fact that the locations of the lamps are not symmetrical (along $y = 2$ m and $y = 5$ m lines). From Figure 2.18b, it is clear that a BER $< 1.2 \times 10^{-8}$ can be achieved over the entire room, and a BER $<10^{-9}$ performance is achieved in most parts of the room except for those directly under lamps. Therefore, high-speed OWC with satisfying performance and mobility can be provided over the entire room.

In addition to the SNR and BER performances, based on the established system model, the signal multipath dispersion can also be simulated. This is achieved by considering the reflections of signal light. Due to the use of LOS link where the transmitter is located at the ceiling and the receiver is located at the CF, the first reflection occurs when the signal light strikes to the floor. The reflected signals will then be reflected again by the walls and the ceiling. After multiple reflections, those arriving at the receiver with an incident angle smaller than the receiver FOV result in the multipath dispersion. Based on the ray-tracking algorithm, we simulate the channel impulse response with 1 ns signal pulse and when the receiver is located at one of the room corners (0, 0, 1). The simulation result is shown in Figure 2.19. We consider the room corner location since it is the most vulnerable one to multipath

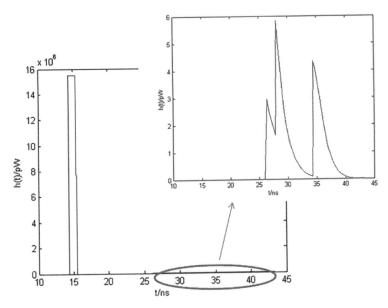

FIGURE 2.19
Simulated impulse response which shows negligible multipath dispersion.

dispersion due to the vicinity of strong reflecting walls. It is clear from the figure that the strength of the signal reflected and captured by the receiver is more than 60 dB lower than that of the signal from the direct LOS channel. Therefore, the multipath dispersion in the system example is very weak and can be discarded.

2.3.5 High-Speed Near-Infrared Indoor OWC System Demonstration

The high-speed near-infrared indoor OWC system example discussed, modeled and analyzed in the previous sections has been experimentally demonstrated as well. The demonstration experiment set-up is shown in Figure 2.20.

In the experiment, a DFB laser is used as the light source, and the wavelength is selected at 1547.15 nm. The 1550 nm wavelength region is selected due to three major reasons: (1) the free-space propagation loss is low; (2) the ambient light intensity is low; and (3) lasers in this band are widely available at low cost since they are mature and widely used in optical fiber communications. The light source is externally modulated using a Mach–Zehnder modulator (MZM) by the data to be transmitted, which is emulated by 2^{31}-1 PRBS (i.e. pseudo-random binary sequence), and the OOK modulation format is used. Since MZM is generally polarization-dependent, a polarization controller (PC) is inserted between the DFB laser and the MZM.

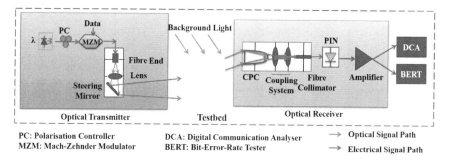

FIGURE 2.20
The high-speed indoor OWC system demonstration experiment set-up.

In addition to the external modulation used here, direct modulation can also be used, where the DFB laser driving current changes according to the data to be transmitted. Because of the use of external modulation, the modulated light is radiated into the free space from an optical fiber end. The standard single-mode fiber is used here (Corning SMF-28), and the radiation beam pattern can be reasonably approximated as Gaussian, with the beam waist $\omega_0 = 5.2\,\mu m$ occurring at the output plane of the fiber [18]. Then this signal passes through a lens to control its divergence before propagating in the free space. A MEMS-based steering mirror adaptively changes the beam propagation direction to cover the user's location and the surrounding area to provide limited mobility.

The LOS link is used in the experiment to achieve high-speed wireless data transmission. After free-space propagation, the signal is filtered by an optical bandpass filter, captured by a CPC and detected by a PD at the receiver side. The PIN PD in the experiment has about 0.8 A/W responsivity at the 1550 nm band. Due to the device limit, in the demonstration a small sensitive area PD is used. Since the PD is smaller than the output surface area of the CPC, a coupling system consisting of multiple lenses and a fiber collimator are used for efficient optical signal coupling. After signal detection, the converted electrical signal is amplified and then measured with a bit-error-rate tester (BERT) and a broadband digital communication analyzer (DCA). In addition to the signal light, background light from direct overhead lamps is also included in the measurements.

When the signal beam covers 1 m area surrounding the user and when the optical power launched into the free space at the transmitter is fixed at 5.5 dBm, the BER performance is shown in Figure 2.21. The OWC bit rate is 2.5 Gbps. As described in the operation principle of the system example, limited mobility is provided to user which enables free user movement inside the signal beam coverage area. Therefore, here we characterize the system BER as a function of the distance from signal beam center. Using DCA, we can also capture the eye-diagram of the

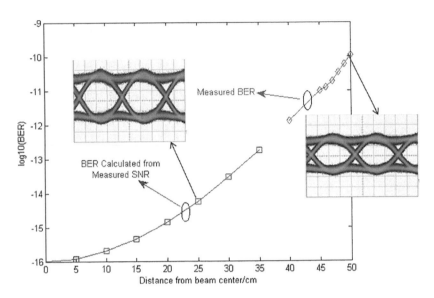

FIGURE 2.21
2.5 Gbps system BER measurement results.

received signal. The eye diagrams when the distance from beam center was 25 and 50 cm, respectively, are also shown in the figure. When the distance from beam center is small, the BER is lower than 10^{-12}, which takes considerably long time to measure. In these cases, the SNR of received signal is measured and the BER is calculated according to Eq. (2.42). It can be seen from the measurement result that the BER increases with the distance from beam center. This is because of the Gaussian intensity distribution at the receiver side, where the collected optical power is lower at locations further away from the beam center. This is also confirmed by the eye diagrams, where the eye opening becomes smaller when the distance from beam center changes from 25 to 50 cm. It is also clear from the figure that a BER $<10^{-9}$ can be achieved over the entire beam coverage area, which can be considered as error-free operation. When combined with the beam steering and user localization functions, high-speed OWC can be achieved over the entire room for personal and local area communication network applications.

In addition to the demonstration experiment discussed previously, even higher speed personal and local area communications using the near-infrared OWC technology has been achieved. Most demonstrations use the LOS link configuration together with beam steering, since the system bandwidth is not limited by multipath dispersion. For example, still using MEMS steering mirrors, up to 12.5 Gbps OWC has been realized [19]. When further combined with the WDM principle, over 50 Gbps (4 × 12.5 Gbps) has been demonstrated [20]. In addition to the MEMS steering mirror, the optical diffractive grating has also been used for the beam steering function, where over 30 Gbps indoor OWC has been reported [21].

2.4 Background Light in Indoor OWC Systems

Both OWC and optical fiber communication systems rely on the use of light to carry and transmit data, and most parts of systems are identical, such as the light source, the modulator and the PD. The major difference is the optical signal transmission medium. The free space is used as the transmission medium in OWC systems. Compared to optical fiber communication systems, in addition to the unavoidable transmission link, the free-space medium also adds background light to the communication system. Equivalently, the background light and the resulting noise are unique in OWC systems. In this section, we investigate the impact of this unique background light.

Here we still focus on the high-speed near-infrared indoor OWC system example studied in Section 2.3, and we still consider the OOK modulation format. As discussed in Section 2.3.3, the signal-dependent noise is normally small and can be neglected. Therefore, the noise variances σ_0^2 and σ_1^2 associated with the transmitted signal "0" and "1" are the same and can be expressed by Eq. (2.37). The main noise sources are the preamplifier-induced noise σ_{pr}^2 and the background light-induced noise σ_{bn}^2. The system SNR can then be calculated using Eq. (2.41). If there is no free-space transmission in the system (i.e. the back-to-back case), there will be no background light-induced noise. Hence, the SRN can be calculated as:

$$\text{SNR} = \left(\frac{R \times (P_{s1} - P_{s0})}{\sigma_0 + \sigma_1}\right)^2 = \left(\frac{R \times (P_{s1} - P_{s0})}{2\sigma_{pr}}\right)^2 \tag{2.43}$$

To achieve the same SNR in the system with free-space transmission, a larger received optical power is required due to the existence of additional background light-induced noise. Here we define the difference in the required received optical power to achieve the same system SNR as the power penalty due to background light-induced noise, and it can be calculated from Eqs. (2.41) and (2.43) as [22]:

$$\text{Power} - \text{Penalty}\,(\text{dB}) = 5 \times \log_{10} \frac{\sigma_{pr}^2 + \sigma_{bn}^2}{\sigma_{pr}^2} \tag{2.44}$$

Normally, we can select the preamplifier bandwidth according to the system bit rate to minimize the induced noise, as shown by Eq. (2.39). This can be done by adding appropriate electrical low-pass filters at the receiver. Therefore, to characterize the background light-induced power penalty, here we study different bit rates of 1, 2.5, 5, and 10 Gbps, using the system model established in Section 2.3.4. We consider background light power ranging from 0 to 10 μW, and the simulation results are plotted in Figure 2.22.

It can be seen from the figure that the power penalty increases with received background light power, whilst it decreases with the bit rate for a fixed received background light power. This is because that according to Eq. (2.39), the preamplifier-induced noise variance increases with the bandwidth of the system. Therefore, in higher speed OWC systems, the preamplifier-induced noise becomes larger and dominates over the background light-induced noise.

To better observe the change of power penalty due to background light with respect to the system bit rate, we show the simulated power penalty for different system speeds when the collected background light power is fixed at −34.5, −30 and −27 dBm, respectively, as shown in Figure 2.23. As shown by Figure 2.17, even when the optical receiver is directly under strong background lamps, the received background light power is still lower than −27 dBm. In this case, it can be seen from Figure 2.23 that the power penalty due to background light-induced noise in the system we studied is always

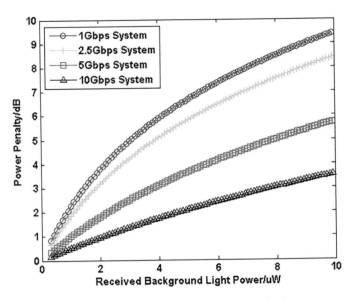

FIGURE 2.22
Simulation result on the power penalty due to background light with respect to the received background light power.

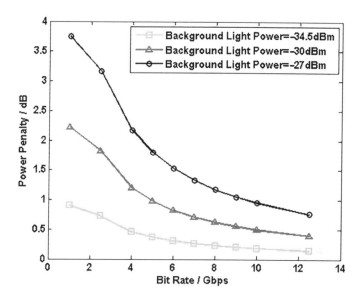

FIGURE 2.23
Simulation results on the power penalty due to background light with respect to the system bit rate.

smaller than 4 dB when the bit rate is <2 Gbps, and it is smaller than 1.5 dB when the bit rate is >7 Gbps. When the overhead lamps are turned off, the received background light power is usually lower than −30 dBm. In this case, the power penalty is reduced to less than 2 dB when the system bit rate is >2 Gbps. In addition, with advanced receivers such as the previously mentioned angle-diversity receiver and imaging receiver, which can reject most of the background light, the received background light power can be further reduced (i.e. the −34.5 dBm case shown in the figure). In this case, the power penalty is lowered to less than 1 dB.

From the earlier simulation results and discussions, it is obvious that for lower bit rate (<5 Gbps) indoor OWC systems, the noise is dominated by the background light-induced noise. However, the power penalty is generally <4 dB. Therefore, using OWC systems to achieve high-speed personal and LANs is feasible with proper power budget.

2.5 Conclusions

In this chapter, we have studied the typical link configurations in high-speed near-infrared indoor OWC systems, including both the LOS- and the non-LOS-based systems. We have reviewed the major parts of OWC systems – the optical transmitter, the free-space link and the optical receiver. For the optical transmitter, both laser and LEDs can serve as the light source; and for the receiver, both PIN and APD can be used. We have also studied the theoretical modeling of the transmitter, the link and the receiver. Normally the light pattern radiated by LED can be modeled as the generalized Lambertian profile, and the light emitted by the laser has Gaussian profile. Depending on the link configuration, the optical signal after free-space propagation can be modeled directly or using the ray-tracing method. Upon light detection, the major noise sources, which are the preamplifier-induced noise and the background light-induced noise, have been investigated.

We have also applied the theoretical models developed for each part in a high-speed near-infrared indoor OWC system example to establish the system model. The LOS link together with beam steering is used in the system example. Based on the system model, the background light, SNR and BER have been simulated and analyzed. Compared to conventional optical fiber communication systems, the background light and its induced noise are unique in OWC system. Therefore, we have carried out detailed analysis to investigate the power penalty due to the background light based on the system example. Results have shown that the background light induces larger power penalty in lower speed systems.

The high-speed near-infrared indoor OWC system example investigated in detail in this chapter is based on the LOS link configuration. It should be noted that up to 10 Gbps high-speed OWC has also been achieved using the non-LOS link [23], where adaptive beam angle and power allocation and diversity receivers are employed.

References

1. J.M. Kahn and J.R. Barry, Wireless infrared communications. *Proceedings of IEEE*, 1997. **85**(2): pp. 265–298.
2. T. Ozugur, J.A. Copeland, M. Naghshineh, and P. Kermani, Next-generation indoor infrared LANs: Issues and approaches. *IEEE Personal Communications*, 1999. **6**(6): pp. 6–19.
3. C.W. Oh, E. Tangdiongga, and A.M.J. Koonen, Steerable pencil beams for multi-Gbps indoor optical wireless communication. *Optics Letter*, 2014. **39**(18): pp. 5427–5430.
4. K. Wang, A. Nirmalathas, C. Lim, and E. Skafidas, High-speed optical wireless communication system for indoor applications. *IEEE Photonics Technology Letters*, 2011. **23**(8): pp. 519–521.
5. G.W. Marsh, and J.M. Kahn, 50-Mb/s diffuse infrared free-space link using on-off keying with decision feedback equalization. *IEEE Photonics Technology Letters*, 1994. **6**(10): pp. 1268–1270.

6. J.M. Kahn, W.J. Krause and J.B. Carruthers, Experimental characterization of nondirected indoor infrared channels. *IEEE Transactions on Communications*, 1995. **43**: pp. 1613–1623.

7. J.B. Carruther and J.M. Kahn, Angle diversity for nondirected wireless infrared communication. *IEEE Transactions on Communications*, 2000. **48**(6): pp. 960–969.

8. G. Yun and M. Kavehrad, Spot diffusing and fly-eye receivers for indoor infrared wireless communications, in *IEEE Conference on Selected Topics in Wireless Communications*, IEEE, pp. 262–265, 1992.

9. S. Jivkova and M. Kavehrad, Indoor wireless infrared local access, multi-spot diffusing with computer generated holographic beamsplitters, in *IEEE International Conference on Communications*, Vancouver, BC, 1999.

10. A.G., Al Ghamdi and J.M.H. Elmirghani, Analysis of diffuse optical wireless channels employing spot-diffusing techniques, diversity receivers, and combining schemes. *IEEE Transactions on Communications*, 2004. **52**(10): pp. 1622–1631.

11. A.G. Al-Ghamdi and J.M.H Elmirghani, Line strip spot-diffusing transmitter configuration for optical wireless systems influenced by background noise and multipath dispersion. *IEEE Transactions on Communications*, 2004. **52**(1): pp. 37–45.

12. J.R. Barry, J.M. Kahn, W.J. Krause, E.A. Lee, and D.G. Messerschmitt, Simulation of multipath impulse response for indoor wireless optical channels. *IEEE Journal on Selected Areas in Communications*, 1993. **11**: pp. 367–379.

13. K. Wang, A. Nirmalathas, C. Lim, and E. Skafidas, High-speed duplex optical wireless communication system for indoor personal area networks. *Optics Express*, 2010. **18**(24): pp. 52199–25216.

14. F. Khozeimeh and S. Hranilovic, Dynamic spot diffusing configuration for indoor optical wireless access. *IEEE Transactions on Communications*, 2009. **57**(6): pp. 1765–1775.

15. B. Leskovar, Optical receivers for wide band data transmission systems. *IEEE Transactions on Nuclear Science*, 1989. **36**(1): pp. 787–793.

16. F.E. Alsaadi and J.M. Elmirghani, Adaptive mobile line strip multibeam MC-CDMA optical wireless system employing imaging detection in a real indoor environment. *IEEE Journal of Selected Areas on Communications*, 2009. **7**(9): pp. 1663–1675.

17. J.B. Carruthers, *Multipath Channels in Wireless Infrared Communications: Modeling, Angle Diversity and Estimation*, University of California, Berkeley, CA, 1997.

18. P. LoPresti, H. Refai, and J. Sluss, Adaptive power and divergence to improve airborne networking and communications, in *24th Digital Avionics Systems Conference*, Washington, DC, 2005.

19. K. Wang, A. Nirmalathas, C. Lim, and E. Skafidas, Ultra-broadband indoor optical wireless communication system with multimode fiber. *Optics Letters*, 2012. **37**(9): pp. 1514–1516.

20. K. Wang, A. Nirmalathas, C. Lim, and S. Skafidas, High speed 4×12.5 Gbps WDM optical wireless communication systems for indoor applications, in *Optical Fiber Communication Conference*, OSA, Washington, DC, 2011.

21. T. Koonen, J. Oh, K. Mekonnen, Z. Cao, and E. Tangdiongga, Ultra-high capacity indoor optical wireless communication using 2D-steered pencil beams. *Journal of Lightwave Technology*, 2016. **34**(20): pp. 4802–4809.

22. K. Wang, A. Nirmalathas, C. Lim, and E. Skafidas, Impact of background light induced shot noise in high-speed full-duplex indoor optical wireless communication systems. *Optics Express*, 2011. **19**(22): pp. 21321–21332.
23. M.T. Alresheedi and J.M. Elmirghani, Performance evaluation of 5 Gbit/s and 10 Gbit/s mobile optical wireless systems employing beam angle and power adaptation with diversity receivers. *IEEE Journal of Selected Areas on Communications*, 2011. **29**(6): pp. 1328–1340.

3

Spatial Diversity Techniques in Near-Infrared Indoor Optical Wireless Communication (OWC) Systems

High-speed near-infrared indoor OWC systems have been demonstrated using both LOS and non-LOS link configurations. Since the non-LOS link-based system utilizes the diffusive reflected signals to cover the entire indoor area, such as an office, it has the advantages of supporting full user mobility and providing the robustness against physical shadowing. However, very high transmission power is required, resulting in eye and skin safety concerns. In addition, to achieve over Gbps communications, advanced receivers (e.g. angle diversity receiver) and adaptive control technologies (e.g. adaptive power allocation) are compulsory, leading to significantly increased system complexity [1]. Therefore, the LOS link is preferred in over Gbps indoor OWC systems.

The LOS link-based indoor OWC system has two key advantages: (1) ultra-high channel bandwidth and (2) high energy efficiency with low transmission power. However, it is also limited by two key factors: first, it is challenging to provide mobility to users and second, the system is vulnerable to physical shadowing, which results in link blockage and communication interruption. The mobility limit can be solved by using the beam steering principle combined with the user localization, which tracks the user movement and adjusts the signal beam direction to maintain high-speed optical wireless data transmission [2]. However, this solution does not improve the system's robustness against physical shadowing.

To solve the vulnerability to physical shadowing in the indoor OWC system with LOS link, the spatial diversity principle can be used. The spatial diversity can be used at the transmitter side, the receiver side or at both ends. Since the optical receiver is located at the user side in this type of systems, low cost and low complexity are typically preferred. Therefore, it is more practical to implement transmitter diversity in the indoor OWC system. In this chapter, we will investigate the spatial diversity techniques for indoor near-infrared OWC systems, focusing on the transmitter diversity techniques. This chapter is organized as follows: Section 3.1 provides the basics of spatial diversity principles and techniques; Sections 3.2 describes the STBC and the RC techniques; Section 3.3 provides a detailed comparison of STBC and RC; Section 3.4 presents a novel delay-tolerant spatial diversity technique which does not require the perfect synchronization between multiple channels; and Section 3.5 concludes this chapter.

3.1 Spatial Diversity Techniques in Indoor OWC Systems

In the LOS link-based indoor OWC systems, the communication link can be blocked by physical shadowing, such as cubicle partitions, furniture or moving personnel, especially when the user is moving. Such blockage results in the unavailability of LOS link and the service interruption. To solve this issue, as shown in Figure 3.1, multiple transmitters can be used to provide multiple LOS links to the user. This approach is called the transmitter diversity technique, which provides redundant OWC links, and it is one widely used spatial diversity technique in wireless communications. By appropriately designing the locations of multiple transmitters, the chance that all channels are blocked simultaneously can be minimized to almost zero, and hence, better system robustness can be achieved.

In addition to the transmitter diversity, multiple receivers can also be used in the system, and this is the receiver diversity technique. Depending on if one or more transmitter(s)/receiver(s) are used, the spatial diversity configuration can be divided into three categories: (1) multi-input single-output (MISO) which uses the transmitter diversity; (2) single-input multi-output (SIMO) which uses the receiver diversity and (3) multi-input multi-output (MIMO), which uses both transmitter and receiver diversities. The typical configuration of three types of systems is shown in Figure 3.2. The system configuration studied in Chapter 2 is also shown, which is also referred to as the single-input single-output (SISO) system without any diversity technique.

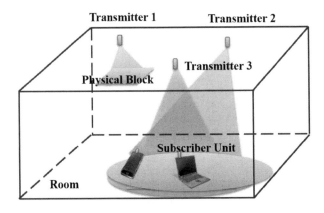

FIGURE 3.1
Indoor OWC system with multiple transmitters and channel blockage.

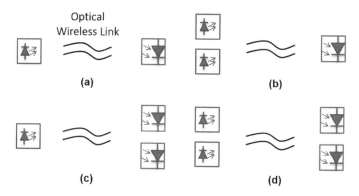

FIGURE 3.2
Spatial diversity configurations in indoor OWC systems. (a) Without spatial diversity (SISO); (b) with transmitter diversity (MISO); (c) with receiver diversity (SIMO) and (d) with both transmitter and receiver diversities (MIMO).

The indoor OWC systems with spatial diversity can be characterized by the channel matrix H, which can be written as

$$H = \begin{bmatrix} h_{11} & \cdots & h_{1n} \\ \vdots & \ddots & \vdots \\ h_{m1} & \cdots & h_{mn} \end{bmatrix} \tag{3.1}$$

where h_{ij} is the channel impulse response between the transmitter i $(i = 1, 2, \ldots, m)$ and the receiver j $(j = 1, 2, \ldots, n)$. Using the channel response matrix, the received signal y can then be calculated as

$$y = Hx + n \tag{3.2}$$

where x is the transmitted signal by multiple optical transmitters and n is the additive channel noise collected by the receivers.

When receiver diversity is used, each optical receiver converts the collected light signal into the electrical domain. The converted signals need to be combined before further processing. A number of combining schemes have been developed, and the most widely used ones include the equal gain combining (EGC), the optimal combining (OC), the maximum ratio combining (MRC) and the switch and examine combining (SEC). In indoor OWC systems, the optical receiver is equipped to each user, and hence, it needs to be simple and low-cost. On the other hand, the transmitter is located at the ceiling, and it serves multiple users. Hence, the cost and complexity of the transmitter are shared by multiple users. Therefore, compared to the receiver diversity, the transmitter diversity technique is more widely used due to the practical application consideration. In this book, we will mainly focus on the transmitter diversity. Readers interested in the receiver diversity technique can refer to [3], where detailed analysis and discussion of the receiver diversity technique are presented.

3.2 High-Speed Indoor OWC System with Transmitter Diversity

In indoor OWC systems with transmitter diversity, multiple optical transmitters available can be used to transmit the same signal or different signals, and multiple diversity coding schemes can be used. Two most widely used transmitter diversity schemes in indoor applications are the STBC and RC. Different signals are transmitted in STBC, and the same signal is transmitted when RC is used. We will discuss these schemes in this section.

3.2.1 Repetition Coded Indoor OWC System

The general principle of RC in typical indoor OWC systems is shown in Figure 3.3, where we consider the case with two transmitters for simplicity. The following analysis and discussion can be easily extended to systems with more transmitters. In the RC scheme, identical signals are sent from both transmitters in each bit interval. For example, the signal s_1 is sent by both transmitters at time slot T_1 and the signal s_2 is sent by both transmitters at the next time slot T_2. These two signal streams are generated with the repetition encoder, and they then modulate two optical transmitters Tx_1 and Tx_2. The structure of each optical transmitter can be the same as that shown in Figure 2.20, which consists of the light source (i.e. laser), the MZM, the lens and the steering mirror. The optical transmitters launch modulated light signals into the free space, and after wireless transmission, the two optical signals are collected and detected by the same optical receiver, which consists

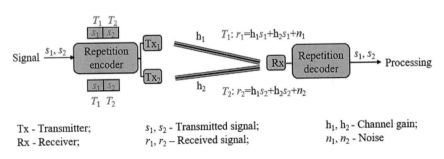

Tx - Transmitter; s_1, s_2 - Transmitted signal; h_1, h_2 - Channel gain;
Rx - Receiver; r_1, r_2 – Received signal; n_1, n_2 - Noise

FIGURE 3.3
The RC principle in indoor OWC systems.

of the bandpass filter, the optical concentrator and the PD. If the two optical channels are well synchronized, the two signals transmitted via the optical wireless link are linearly combined at the receiver. Therefore, the received optical signal during the time slot t can be described as

$$Rx_t = \sum_{i=1}^{2} h_{i,t} s_t + n_{ot} \tag{3.3}$$

where $h_{i,t}$ is the channel impulse response from Tx_i to the receiver during time slot t, s_t is the transmitted signal and n_{ot} is channel noise in the optical domain. The collected optical signal is then converted to photocurrent I_{Rx} by the PD. Assume the average transmission power of the two transmitters is $\langle P_i^{tx} \rangle$ ($i = 1$ or 2), then the photocurrent can be expressed as

$$I_{Rx} = R_d \times \sum_{1}^{2} h_i P_i^{tx} + n \tag{3.4}$$

where R_d is the responsivity of PD and n is electrical noise at the receiver, which can be modeled as the additive white Gaussian noise (AWGN) with main contributions from shot and thermal noises. Here we assume that the channel impulse response during the time period considered is constant, which is a reasonable assumption due to the slow channel change (e.g., the low walking speed of moving users) and the high data transmission speed in indoor applications. The generated photocurrent in the RC-based indoor OWC system is then converted to a voltage signal by a TIA, and the signal is then decoded and further processed.

As detailed in Section 2.3.3, the major noise sources in the system are the pre-amplifier-induced noise σ_{pr}^2 and the background light-induced noise σ_{bn}^2. Using Eqs. (2.39) and (2.40), the SNR of the RC-based system can be calculated as [4]

$$\text{SNR} = \frac{\left(R_d \times \sum_{1}^{2} h_i \langle P_i^{tx} \rangle \right)^2}{\sigma_{pr}^2 + \sigma_{bn}^2} \cong \frac{\left(R_d \times \sum_{1}^{2} h_i \langle P_i^{tx} \rangle \right)^2}{2qR_d \times \left(\sum_{1}^{2} h_i \langle P_i^{tx} \rangle + P_{bn} \right) I_2 \Delta f + \dfrac{4k_B T}{R_F} I_2 \Delta f} \tag{3.5}$$

where q is the electron charge, P_{bn} is the collected background light power, I_2 is the noise bandwidth factor, Δf is the receiver bandwidth, k_B is the Boltzmann's constant, T is the absolute temperature and R_F is the feedback resistance in the TIA. Here we neglect the FET $1/f$ noise and the FET white channel noise for simplicity [5], since they are generally much weaker.

If the simplest OOK modulation is used in the RC-based indoor OWC system, the BER performance can then be calculated as [4,6]

$$\text{BER}_{RC-OOK} = Q\left(\sqrt{\text{SNR}}\right) \tag{3.6}$$

where $Q(x) = \dfrac{1}{\sqrt{2\pi}} \displaystyle\int_{x}^{\infty} e^{-\frac{y^2}{2}} \, dy$.

3.2.2 Space-Time Coded Indoor OWC System

In addition to the RC-based transmitter diversity scheme, the STBC is also widely considered in indoor OWC systems. The general principle of STBC-based system with two transmitters is shown in Figure 3.4. In STBC, signals transmitted by the two transmitters are different, and two adjacent time slots are used together. The transmission signal matrix can be described as

$$S = \begin{bmatrix} s_1 & s_2 \\ \sim s_2 & s_1 \end{bmatrix} \tag{3.7}$$

where the column of the matrix represents the transmitter Tx_i ($i = 1$ or 2), the row represents the time slot T_j ($j = 1$ or 2) and $\sim s$ represents the bit-wise not of the original signal, where $\sim s = 1 - s$. Due to the typical use of intensity modulation and direct detection (IM/DD) in indoor OWC systems, here we employ the extended Alamouti type of STBC, where all signals have non-negative real-values.

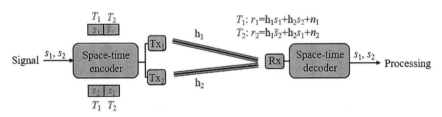

FIGURE 3.4
The STBC principle in indoor OWC systems.

After free-space propagation, the signals with STBC are collected and detected by the optical receiver. The received optical signal during the time slot j can be expressed as

$$Rx_j = \sum_{i=1}^{2} h_{i,j} s_{i,j} + n_{o,j} = \sum_{i=1}^{2} h_i s_{i,j} + n_o, \quad j = 1 \text{ or } 2 \tag{3.8}$$

where $s_{i,j}$ is the signal transmitted by the transmitter Tx_i during time slot T_j, $h_{i,j}$ is the channel impulse response between the transmitter Tx_i and the receiver during time slot T_j, and $n_{o,j}$ is the channel noise in the optical domain during time slot T_j. Due to the slow-change characteristic of indoor optical wireless channels, both the channel impulse responses and the optical noise can be considered as constant during adjacent time slots.

If we assume that the transmission power matrix corresponding to the signal matrix S is $\begin{pmatrix} \langle P_{Tx1\_T1} \rangle & \langle P_{Tx2\_T1} \rangle \\ \langle P_{Tx1\_T2} \rangle & \langle P_{Tx2\_T2} \rangle \end{pmatrix}$, the generated photocurrent vector by the PD in two consecutive bit intervals can be calculated as

$$\vec{I}_{Rx} = R_d \times \begin{pmatrix} \sum_1^2 h_i \langle P_{Txi\_T1} \rangle \\ \sum_1^2 h_i \langle P_{Txi\_T2} \rangle \end{pmatrix} + \vec{n} \tag{3.9}$$

where $\vec{n}$ is the noise vector. In STBC-based indoor OWC systems, the maximum likelihood (ML) detection is usually employed. Hence, if we assume that the transmission power of each transmitter during adjacent time slots remains unchanged $\left(\text{i.e.} \langle P_{Tx1\_T1} \rangle = \langle P_{Tx1\_T2} \rangle \text{ and } \langle P_{Tx2\_T1} \rangle = \langle P_{Tx2\_T2} \rangle \right)$, the SNR at the receiver after the PD can be described by

$$\text{SNR} = \frac{\left(R_d \times \sqrt{\sum_1^2 \left(h_i \langle P_i^{tx} \rangle \right)^2} \right)^2}{\sigma_{\text{shot}}^2 + \sigma_{\text{thermal}}^2}$$

$$\cong \frac{\left(R_d \times \sqrt{\sum_1^2 \left(h_i \langle P_i^{tx} \rangle \right)^2} \right)^2}{2qR_d \times \left(\sqrt{\sum_1^2 \left(h_i \langle P_i^{tx} \rangle \right)^2} + P_{bn} \right) I_2 \Delta f + \dfrac{4k_B T}{R_L} I_2 \Delta f} \tag{3.10}$$

where $\langle P_i^{tx} \rangle$ ($i = 1$ or 2) is the average transmission power in the corresponding channel. If the OOK modulation format is used in the system, the conditional BER for STBC can then be calculated by Eq. (3.6).

3.3 Comparison of STBC- and RC-Based High-Speed Indoor OWC System

The basic principles of RC and STBC in near-infrared indoor OWC systems are discussed in Section 3.2. Both types of spatial diversity schemes use multiple transmitters for the signal transmission to provide spatial redundancy. In the system example with two transmitters, when one of the OWC channels is blocked by physical shadowing, the other channel can still maintain high-speed wireless connectivity, and hence, service interruptions of users will be significantly reduced. When both OWC channels are available, each transmitter can be operated at the maximum allowed power level (limited by safety regulations). In this case, a larger total power is available at the receiver side, which can be used to either extend the beam coverage area or to increase the communication speed.

The comparison of RC and STBC has been widely studied in RF systems. Results have shown that in principle, STBC can provide better system performance (e.g. BER or capacity) than RC, since full diversity can be achieved in STBC. It is important to note that the coherent detection is normally used in RF systems, where in addition to the amplitude information, the phase information can also be retrieved for signal detection. On the other hand, indoor OWC systems normally employ the direct detection, where a PD is used to convert the input optical signal into the corresponding electrical signal (i.e. photocurrent). Therefore, only power information can be detected whilst the phase information is not available. As a result, the comparison of RC and STBC in RF systems is no longer valid in indoor OWC systems.

In this section, we conduct a detailed comparison of RC and STBC in indoor OWC systems with direct detection. The comparison is based on the theoretical framework of RC and STBC as described by Eqs. (3.3)–(3.10). Using the equations, the BER performance of both RC and STBC schemes in high-speed near-infrared indoor OWC systems is numerically simulated. The simulation results are shown in Figure 3.5. The received optical power shown in the figure is $\sum_1^2 h_i \langle P_i^{tx} \rangle$, which is the total received power from both transmitters. For simplicity, we assume that the received optical powers from two OWC channels are identical $\left(\text{i.e. } h_i \langle P_1^{tx} \rangle = h_i \langle P_2^{tx} \rangle\right)$. We also assume that an optical bandpass filter with relatively narrow bandwidth is used at the receiver, and hence, the collected background light power is reduced to a negligible level (i.e. $P_{bn} \approx 0$). The PIN PD is used, and it has a responsivity $R_d = 0.8$ A / W. The transmission data rate is assumed to be 10 Gb/s with the OOK modulation format.

It can be seen from the simulation results in Figure 3.5 that different from RF systems, the indoor OWC system with RC always outperforms the system with STBC. When the forward error correction (FEC) code is used in the system, a BER performance of better than 10^{-3} can be considered as error-free. Therefore, here we characterize the receiver sensitivity

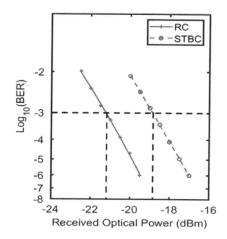

FIGURE 3.5
Simulated BER performance comparison between RC and STBC.

at the BER of 10^{-3}. It can be seen from the figure that the receiver sensitivity of the systems with RC and STBC are about −21.2 and −18.9 dBm, respectively. Therefore, the RC scheme can achieve more than 2 dB receiver sensitivity improvement in indoor OWC systems with direct detection. This observation is consistent with the theoretical analysis shown in Eqs. (3.5) and (3.10). From Eq. (3.5), it is clear that the SNR of RC-based system is proportional to $\left(\sum_i^2 h_i \langle P_i^{tx} \rangle \right)^2$, whilst from Eq. (3.10), it can be seen that the SNR of STBC-based system is proportional to $\left(\sqrt{\sum_i^2 (h_i \langle P_i^{tx} \rangle)} \right)^2$. Since $\sum_i^2 h_i \langle P_i^{tx} \rangle > \sqrt{\sum_i^2 (h_i \langle P_i^{tx} \rangle)^2}$ and both systems have the same amount of noise, the RC-based indoor OWC system requires a lower received power level to achieve the same SNR. Therefore, a better BER can be achieved in the RC scheme compared to the STBC scheme.

The indoor OWC systems with RC and STBC have also been experimentally investigated. The experimental setup for the comparison is shown in Figure 3.6. Here, the OOK modulation format is used, and the bit rate of each channel is 10 Gbps. The modulated data is then encoded using RC or STBC as described in the previous section. An arbitrary waveform generator (AWG) is used to complete the data generation, modulation and transmitter diversity encoding offline. Then the two signals are used to modulate two laser sources using external MZMs. The operation wavelengths are 1550 and 1549 nm, respectively. The two lasers can be operated at the same wavelength as well, and here we use two wavelengths for operation simplicity consideration. An optical attenuator is employed after each MZM to control the transmitted optical power. We use this controlled power attenuation to emulate additional signal power losses due to the optical beam blockage in OWC channels. An optical delay line (ODL) is also inserted in one of the two links to introduce controllable optical path delay. The optical path delay

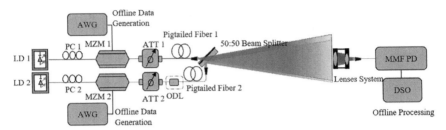

AWG - Arbitrary Waveform Generator; LD: laser; PC - Polarization Controller; MZM - Mach Zehnder Modulator; ATT - Optical Attenuator; ODL - Optical Delay Line; MMF - Multimode Fiber; PD - Photodetector; DSO - Digital Storage Oscilloscope

FIGURE 3.6
Experimental setup for investigating indoor OWC systems with RC and STBC spatial diversity schemes.

is highly likely to exist in practical indoor OWC systems with transmitter diversity, and it is mainly caused by the optical wireless channel path length difference. To launch the optical signals into the free space, pigtailed optical fibers are used. The optical signals emitting to the free space have a beam divergence of about 16°. The maximum output power from each transmitter is 4 dBm, and hence, the maximum power level of the combined two beams is still within the laser eye and skin safety limit.

In the experimental demonstration system, the two OWC beams are spatially close to each other, and the free-space signal transmission distance is 1 m. Such transmitter and channel configuration ensures that both beams can be efficiently focused by a specially designed lenses system and then detected by a MMF-coupled PD. The lens is coated with an optical bandpass filter to reject most of the out-of-band background light, which effectively reduces the overall receiver noise. After the PD, the converted electrical signal is sampled by a digital signal oscilloscope (DSO) for offline signal processing.

The measured BER results with only one OWC channel activated are shown in Figure 3.7a as a benchmark. This measurement is taken without any optical beam blockage in the OWC system. Therefore, the two optical wireless channels from the two transmitters have almost identical channel gains are similar, i.e. $h_1 \cong h_2$. This is confirmed by the results shown in the figure, where the two channels have similar BER performance.

Then we measure the system BER performance with RC and STBC, and the results are shown in Figure 3.7b. Here we consider two scenarios. The first scenario is that there is no link blockage in the system, and hence, the two channels in both RC and STBC transmitter diversity schemes have the same channel gain. The BER results in this scenario is shown by the two solid lines. It can be seen that the RC scheme achieves a receiver sensitivity similar with the single-channel benchmark, whilst the receiver sensitivity of the STBC scheme is degraded by about 1.8 dB.

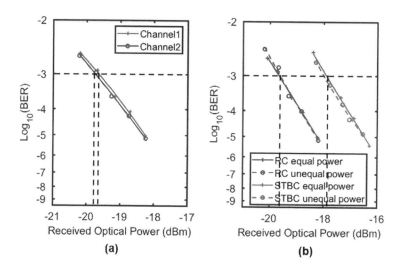

FIGURE 3.7
(a) BER results when only one channel activated and (b) BER results for the RC and STBC schemes.

The second scenario considered here is that one of the two OWC channels is blocked. When the OWC channel is blocked by physical shadowing, the channel gain is affected (i.e. reduced). Hence, in the experiment, we emulate the channel blockage by changing the channel gain of one OWC link. It is worth noting that in our experiment, we emulate the reduced channel gain by introducing additional optical power attenuation in the fiber link through the optical attenuator, which is equivalent to the additional power loss in the free space. We use this approach since it is easier to preciously control the additional loss induced. In this case, the two OWC channels have different channel gains (i.e. $h_1 \neq h_2$). The difference between channel gains in the measurement is selected at 3 dB, and the BER results are shown by the dashed lines in Figure 3.7b. It can be seen from the figure that even with channel blockage, similar performance as that in the first scenario is still achieved, where the RC schemes achieved about 1.8 dB better receiver sensitivity than the STBC scheme. These agree well with the theoretically predicted and simulated results shown in Figure 3.5, confirming that in indoor OWC systems with direct detection, the RC transmitter schemes are better situated.

In indoor near-infrared OWC systems, since mobility is normally provided to end users, it is highly likely that the two optical channels have different optical path lengths. This results in the relative optical delay of the signals sent by two transmitters and imperfect synchronization at the receiver. Normally, channel training is conducted to estimate the channel delay and then to synchronize the OWC channels. However, since the target system data rate is high (over Gbps), it is challenging to achieve perfect

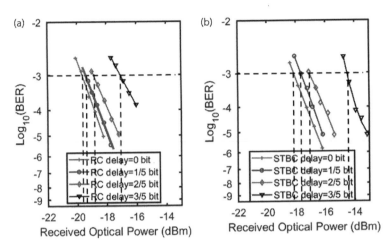

FIGURE 3.8
BER results of transmitter diversity schemes with optical delay. (a) RC scheme and (b) STBC scheme.

synchronization. Therefore, here we investigate the impact of channel delay on the RC and STBC transmitter diversity schemes. We focus on the less than one-bit interval optical delay, since it is more challenging to deal with using the channel training and synchronization process. The results are shown in Figure 3.8.

In the experiment, the optical delay within one-bit interval is realized by changing the controllable ODL inserted at one of the transmitters. The transmitted data rate was kept at 10 Gb/s with OOK modulation. It can be seen from the figure that when the optical delay is within 3/5-bit interval, both RC and STBC schemes can achieve error-free operations in general (with FEC). When the channel delay increases, the BER becomes worse for both transmitter diversity schemes. The BER deterioration speed with respect to the received optical power is faster at higher optical delays. This is because for the OOK modulation without any guard interval, the signals from two transmitters have temporal overlapping at the receiver side, which results in ISI and leads to BER degradation. When the optical delay increases, the overlapping becomes larger and the ISI is more severe, resulting in a faster BER deterioration. In addition, when the delay between two channels increases, it can be seen that the BER performance degradation of the STBC scheme is slightly faster than that of the RC scheme. This is due to the coding and decoding principles of the real-valued extended Alamouti-type STBC, where three possible signal levels exist upon detection and decision, compared to the two possible signal levels in the RC scheme. Therefore, the probability of decision error is larger in the STBC scheme with delayed signals.

Form the above theoretical analysis, simulation and measurement results, we can reach the general conclusion that in indoor OWC systems with direct detection, the RC-based transmitter diversity scheme has better BER performance than the STBC scheme. The RC scheme is also more tolerant to the possible channel delay due to different path lengths.

3.4 Delay-Tolerant High-Speed Indoor OWC System with Transmitter Diversity

From the experimental analysis discussed in Section 3.3, it is clear that both RC and STBC transmitter diversity schemes are vulnerable to the possible channel delay. When the channel delay increases, the BER increases rapidly. However, in high-speed indoor OWC systems with user mobility, the channel delay is highly likely to exist. Generally, to overcome the limit of channel delay, the channel training, delay estimation and synchronization process is utilized. The channel delay can also be solved by using channel training, adding cyclic prefix (CP) and applying signal equalization techniques. However, both approaches reduce the effective system data rate and communication capacity, especially when the channel delay is long when long CP is used (since the CP needs to be longer than the channel delay) and when frequent synchronization is needed. In addition, when the system operation speed is high (e.g. over Gbps), it becomes more difficult to achieve perfect synchronization.

To overcome the channel delay issue when transmitter diversity is used, here we introduce a novel delay-tolerant scheme [7]. Here we select the RC-based transmitter diversity scheme, since it can provide better performance in indoor OWC systems when the direct detection is used. It is worth noting that the delay-tolerant scheme to be discussed in this section is also applicable to the STBC scheme.

The delay-tolerant RC-based indoor OWC system principle is shown in Figure 3.9. For simplicity, we still consider an indoor OWC system with two transmitters and one receiver. For clarity, we show the building blocks of conventional RC-based transmitter diversity scheme in boxes with darker color and the additional building blocks of the proposed delay-tolerant RC-based transmitter diversity scheme in boxed with lighter color.

In the conventional RC scheme, an identical signal $s(t)$ is sent by both transmitters simultaneously. These signals are then collected by one receiver after free-space propagation. In such case, synchronization is required since non-synchronized channels with >1 symbol delay causes overlapping of signals at the receiver, which results in ISI and leads to RC decoding errors. To solve this channel delay limitation, orthogonal filters can be used. The orthogonal filters encode the signals transmitted by difference transmitters, and hence,

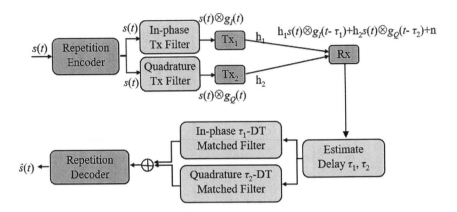

Tx/Rx - Transmitter/Receiver; $g_I(t)$, $g_Q(t)$ - Orthogonal Filters;
DT - Delay Tolerant; h_1, h_2 - Channel Gain;
$s(t)$ - Transmitted Signal;   n - Noise; $\hat{s}(t)$ - Estimated Signal

FIGURE 3.9
Principle of the delay-tolerant RC-based indoor OWC system. Blocks in boxes with darker color are used in the conventional RC scheme, and blocks in boxes with lighter color are additional in the proposed delay-tolerant RC scheme.

signals from different channels can be distinguished at the receiver side with matched filters and the corresponding optical delay can be removed before RC decoding.

In the example system shown in Figure 3.9 with two transmitters, the form of the orthogonal filters $g_I(t)$ and $g_Q(t)$ is a Hilbert pair, similar to the filters used in the CAP modulation [8]. The orthogonal filters can be expressed as

$$g_I(t) = w(t)\cos(2\pi f_c t) \tag{3.11}$$

$$g_Q(t) = w(t)\sin(2\pi f_c t) \tag{3.12}$$

where $w(t)$ is the square root raised cosine pulse window and f_c is the central frequency. With the orthogonal filters, the two transmitted signals can be represented as

$$Tx_1(t) = s(t) \otimes g_I(t) \tag{3.13}$$

$$Tx_2(t) = s(t) \otimes g_Q(t) \tag{3.14}$$

where $\otimes$ represents the convolution operation. Assume the channel impulse responses (i.e. channel gains) from the two transmitters are

h_1 and h_2, respectively, then the received signal after optical wireless transmission can be expressed as

$$Rx(t) = h_1 s(t - \tau_1) \otimes g_I(t - \tau_1) + h_2 s(t - \tau_2) \otimes g_Q(t - \tau_2) + n \qquad (3.15)$$

where τ_1 and τ_2 are the propagation time delays from transmitters 1 and 2 to the receiver, respectively, and n is the additive noise. To estimate the channel delays, the channel training process can be used and special training sequence can be designed to facilitate the delay estimation process [7]. With the estimated channel delays, matched orthogonal filters (i.e. in-phase and quadrature matched filters) are used at the receiver side. Assume the responses of receiver filters are $g_{I,m}(t)$ and $g_{Q,m}(t)$, respectively, then the resulting signals after passing through the matched filters can be represented as

$$r_I(t) = Rx(t) \otimes g_{I,m}(t - \tau_1) \approx h_1 s(t - \tau_1) + n_1 \qquad (3.16)$$

$$r_Q(t) = Rx(t) \otimes g_{Q,m}(t - \tau_2) \approx h_2 s(t - \tau_2) + n_2 \qquad (3.17)$$

where n_1 and n_2 are the additive noises of the two channels, respectively. The noises of two channels can be considered as similar in typical indoor applications. The filtered signals are then decoded to recover the originally transmitted data.

From the working principles discussed above and as shown by Eqs. (3.11)–(3.17), it can be seen that due to the orthogonality of filters applied in the two channels, although the optical signal after OWC link collected by the receiver still suffers channel delay, its impact can be minimized at the receiver side and the transmitted data can be recovered successfully. Therefore, delay-tolerant RC-based transmitter diversity can be realized.

The delay-tolerant RC-based transmitter diversity scheme for indoor OWC systems has been experimentally demonstrated. The experimental setup is similar to that shown in Figure 3.6. At the transmitter side, 2.5 Gb/s data with 2-pulse amplitude modulation (2-PAM) format was generated. The data was split into two parts for RC, and they passed through the orthogonal filters as described by Eqs. (3.11) and (3.12). The filtered two data streams were then modulated onto the optical carrier and launched into the free space for transmission. An ODL was applied in one path to emulate accurate channel delay due to different optical wireless channel lengths. The RC signals were transmitted over the optical wireless link for about 1.2 m length, and they were then captured by the optical receiver. Channel delays were estimated at the receiver side based on training symbols, and matched filters were applied to minimize the impact of channel delay.

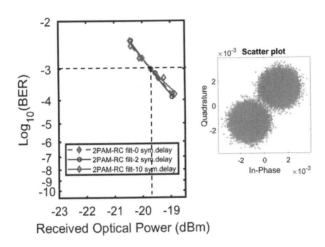

FIGURE 3.10
Measured BER of the 2.5 Gb/s delay-tolerant RC-based indoor OWC system.

The measured BER performance of the proposed delay-tolerant RC-based transmitter diversity scheme is shown in Figure 3.10. No relative channel delay (i.e. $\tau_1 - \tau_2 = 0$), 2 symbol-period relative channel delay and 10 symbol-period relative channel delay cases were measured. The two channel gains h_1 and h_2 were kept the same in the experiment, since such condition results in the largest impact of channel delay. It can be seen from the results that similar BER performance is achieved for all channel delay cases, where the receiver sensitivity (defined at BER = 10^{-3}) is about –19.7 dBm. Therefore, the proposed orthogonal filter-based RC scheme is robust against the channel delay, solving the synchronization issue in indoor OWC systems with transmitter diversity. To further confirm the capability of proposed delay-tolerant RC scheme, the constellation of the received signal after matched filters with 10 symbol-period relative channel delay is also shown in Figure 3.10. It is clear that signals from the two channels (i.e. in-phase and quadrature signals) are orthogonal against each other even with significant channel delay, and hence they can be separated with minimum inter-channel interference.

The discussion and demonstration of the proposed delay-tolerant RC-based transmitter diversity principle above are based on an indoor OWC system with two transmitters. This principle can be extended to more transmitters as well. The general principle is shown in Figure 3.11, where orthogonal filters are employed at each transmitter and corresponding matched filters are used at the receiver side. The higher dimensional orthogonal filters can be designed using the frequency domain filter principle as discussed in [9].

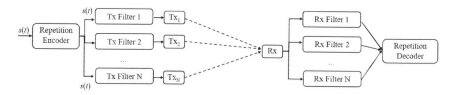

FIGURE 3.11
Delay-tolerant RC-based transmitter diversity principle for more transmitters.

3.5 Conclusions

In this chapter, the diversity techniques in high-speed near-infrared indoor OWC systems have been briefly described and reviewed. The diversity techniques are capable of providing spatial redundancy to improve the system communication performance and to improve the robustness against physical shadowing and channel blockage issues. Both transmitter diversity and receiver diversity schemes have been proposed and studied, and this chapter has focused on the transmitter diversity due to practical system cost and complexity considerations.

A number of transmitter diversity schemes have been proposed and investigated, and we have focused on RC and STBC schemes here due to their popularity. The basic operation principles of both have been discussed in Section 3.2, where RC uses multiple transmitters for sending the same signal and STBC sends different signals through the transmitters using multiple time slots. A comparison of RC- and STBC-based transmitter diversity schemes has been presented in Section 3.3 based on the most widely used IM/DD indoor OWC system, and results have shown that commonly RC can achieve better system performance in terms of receiver sensitivity and communication BER.

Generally, traditional transmitter diversity schemes require signal synchronization and they are vulnerable to channel delays. However, signal synchronization is challenging in high-speed systems and channel delays are highly likely to exist in practical scenarios. To overcome this limit, a delay-tolerant transmitter diversity scheme has been discussed in Section 3.3. In this scheme, orthogonal filters have been applied in multiple wireless channels, and the delay issue has been minimized by using matched filters at the receiver side. Therefore, the vulnerability of transmitter diversity against channel delay has been significantly improved.

In addition to the transmitter diversity principles discussed in this chapter, receiver diversity schemes also have the capability of improving the performance in indoor OWC systems. Details on the receiver diversity can be found in [3].

References

1. M.T. Alresheedi and J.M.H. Elmirghani, Performance evaluation of 5 Gbit/s and 10 Gbit/s mobile optical wireless systems employing beam angle and power adaptation with diversity receivers. *IEEE Journal of Selected Areas on Communications*, 2011. **29**(6): pp. 1328–1340.

2. K. Wang, A. Nirmalathas, C. Lim, and E. Skafidas, High-speed optical wireless communication system for indoor applications. *IEEE Photonics Technology Letters*, 2011. **23**(8): pp. 519–521.

3. Z. Ghassemlooy, W. Popoola, and S. Rajbhandari, Outdoor OWC links with diversity techniques, in *Optical Wireless Communications—System and Channel Modelling with Matlab* CRC Press, Boca Raton, FL, pp. 397–442, 2013.

4. J.M. Kahn and J.R. Barry, Wireless infrared communications. *Proceedings of IEEE*, 1997. **85**(2): pp. 265–298.

5. K. Wang, A. Nirmalathas, C. Lim, and E. Skafidas, Impact of background light induced shot noise in high-speed full-duplex indoor optical wireless communication systems. *Optics Express*, 2011. **19**(22): pp. 21321–21332.

6. M. Safari and M. Uysal, Do we really need OSTBCs for free-space optical communication with direct detection. *IEEE Transactions on Wireless Communications*, 2008. **7**(11): pp. 4445–4448.

7. T. Song, K. Wang, A. Nirmalathas, C. Lim, E. Wong, and K. Alameh, Delay-tolerant repetition-coding for personal-area optical wireless communications with spatial diversity, in *Optical Fiber Communication Conference (OFC)*, San Diego, CA, 2019.

8. L. Tao, Y. Wang, Y. Gao, A. Lau, N. Chi, and C. Lu, Experimental demonstration of 10 Gb/s multi-level carrier-less amplitude and phase modulation for short range optical communication systems. *Optics Express*, 2013. **21**(5): pp. 6459–6465.

9. M.B. Othman, X. Zhang, L. Deng, M. Wieckowski, J.B. Jensen, and I.T. Monroy, Experimental investigations of 3-D-/4-D-CAP modulation with directly modulated VCSELs. *IEEE Photonics Technology Letters*, 2012. **24**(22): pp. 2009–2012.

4

Wavelength Multiplexing and Multi-User Access in Near-Infrared Indoor Optical Wireless Communication Systems

The optical wireless technology transmits data using the optical wave directly through the free space, and ultra-high-speed wireless communications can be realized, especially with the LOS link configuration [1]. In previous chapters, we have discussed the basic fundamentals, system models and spatial diversity principles in indoor OWC systems. Up to 12.5 Gb/s OWC has been demonstrated [2].

To further increase the achievable speed in indoor OWC systems, advanced modulation formats and signal processing algorithms can be used, such as multi-level pulse amplitude modulation (PAM) [3], CAP [4], DMT and OFDM modulation formats [5] and frequency domain or time domain equalizations [6]. However, these approaches usually require more complicated optical wireless transceivers, which increase the cost significantly. In addition, high-speed digital-to-analogue converters (DACs) and analogue-to-digital converters (ADCs) are also needed, which are normally power hungry and expensive.

In addition to using advanced modulation formats and complicated signal processing algorithms, the data rate supported by indoor OWC systems can also be increased by using the wavelength multiplexing technique. In Section 4.1, we will discuss the WDM technique in indoor OWC systems [7,8], where multiple optical carriers with different wavelengths are used together to achieve higher data rates. The possible inter-channel crosstalk between multiple wavelength channels will also be investigated.

In most of previous demonstrations of high-speed indoor OWC systems, the wireless connectivity is only provided to a single user. However, in practical applications in indoor environments, such as in offices and homes, it is critical and vital for indoor OWC systems to support multiple users. In Section 4.2, we will introduce the multi-user access principles, focusing on the traditional frequency-division-multiplexing access (FDMA) [9], code-division-multiplexing access (CDMA) [10] and time-division-multiplexing access (TDMA) [11] schemes, which are widely used in RF systems. We will introduce the basic principles and analyze the major limitations of these schemes.

However, the adoption of traditional multiplexing techniques has several fundamental limitations, such as the high bandwidth requirements of optoelectronic devices and the need for long-length codes. To overcome these issues, we will also introduce the TSC scheme to support multiple users [12] in Section 4.3. We will present an example of indoor OWC system employing the TSC scheme to provide simultaneous access to multiple end users, and we will analyze the impact of the number of users on the system performance. The capability of TSC scheme will also be discussed further. The principle of the TSC scheme with adaptive loading function to improve the spectral efficiency and communication speed will also be introduced [13]. In addition, the imperfect timing issue in the TSC scheme will be analyzed both theoretically and experimentally in this section.

This chapter will then be summarized in Section 4.4.

4.1 Indoor OWC Systems with WDM

In indoor OWC systems, the data-carrying optical signal propagates via the free space for wireless communications. As discussed in Chapter 2, both LOS and diffusive link configurations can be used in indoor OWC systems, and the LOS configuration can achieve higher data rates. Using the direct LOS link, up to 12.5 Gb/s data rate has been demonstrated with the simplest on-off-keying (OOK) modulation format [2]. To further increase the data rate to satisfy the ever-growing speed demand of end users, higher-order advanced modulation formats can be used [3–5]. However, more complicated transceivers, complex signal processing and power-hungry and costly ADCs and DACs are needed, limiting practical applications.

Another approach to increase the achievable data rate in indoor OWC systems is using the multiplexing technique, and one promising solution is using the WDM scheme, which is the dominant technology in current optical fiber communication systems and networks. The general principle of WDM-based indoor OWC system is shown in Figure 4.1.

Inside the OWC transmitter, multiple light sources are used for transmitting different data streams. Either direct modulation or external modulation (e.g. using MZMs) can be used, and these light sources are operated at

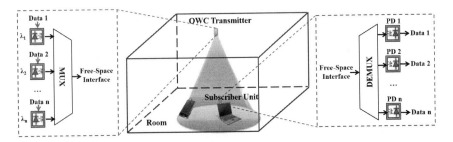

FIGURE 4.1
Indoor OWC system with WDM for higher data rates.

different wavelengths. The modulated light signals are multiplexed using a wavelength multiplexer (MUX) and then launched into the OWC channel with the same free-space interface. One possible structure of the free-space interface is shown in Figure 2.16. The multiplexed optical signals propagate through the free space to the user side, where they are collected by the free-space interface at the subscriber unit. The collected signals are demultiplexed with a wavelength demultiplexer (DEMUX), which separates signals with different wavelengths, and the data streams carried by different wavelengths are detected with separate PDs, which perform the O-E conversion.

In the WMD-based indoor OWC systems, the total data rate is increased by transmitting multiple data streams using different wavelengths simultaneously. Up to 4 × 12.5 Gb/s data rate has been experimentally demonstrated with four wavelength channels and the simple OOK modulation format [7]. Optical signals with different wavelengths are transmitted by the same OWC transmitter and collected by the same OWC receiver, and they propagate through the same OWC link. Therefore, all signals experience similar channel characteristics and have similar performance. However, possible inter-channel crosstalk due to imperfect wavelength filtering at either the transmitter or the receiver side may introduce additional power penalty and limit the system performance, and its impact requires further investigation.

To investigate the impact of inter-channel crosstalk on the performance of WDM-based indoor OWC systems, we can characterize the BER performance of the system when different numbers of wavelength channels are turned ON. We select a total of four wavelength channels, at 1550.12 nm, 1550.92 nm, 1551.72 nm and 1552.52 nm, respectively. Here we select a small spacing of 0.8 nm (i.e. 100 GHz at the 1550 nm band) between adjacent wavelength channels in the investigation for the worst-case scenario, since the possible inter-channel crosstalk is normally inversely proportional to the channel spacing. Such channel spacing is also selected since it is consistent with the WDM standard in optical fiber networks. The data rate of each wavelength channel is 12.5 Gb/s, and the OOK modulation format is selected.

When one wavelength channel, two channels and four channels are turned ON, the BER performance of the WDM-based indoor OWC system is shown in Figure 4.2. The beam footprint at the user side is fixed at 1 m and the BER is characterized with respect to the distance from the center of beam coverage area. Shown in the figure are the results of the 1550.92 nm channel, and the results of other channels are similar. It is obvious that when two or four wavelengths are turned ON, the BER performances are almost the same. The BER performance improves when there is only one channel turned ON in the system. The difference in BER is mainly due to the optical losses incurred by the MUX and the DEMUX, which are used in the systems with multiple wavelength channels. When there is only one wavelength channel in the system, MUX and DEMUX are not inserted in the link. When the losses of MUX and DEMUX are factored into the system with a single wavelength channel, the BER performance is

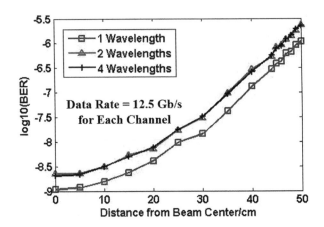

FIGURE 4.2

BER performance of the 1550.92 nm channel with different numbers of wavelength channels turned ON. Channel spacing = 0.8 nm.

similar to that of the system with multiple wavelength channels. Therefore, the power penalty due to the inter-channel crosstalk between different wavelength channels is negligible in the system.

Using the WDM principle, up to 200 Gb/s data rate has been demonstrated in indoor OWC systems with five channels at 40 Gb/s per channel [8]. The power penalty of inter-channel crosstalk has also shown to be negligible, verifying the feasibility of using wavelength multiplexing to increase the achievable communication speed.

In the investigations discussed above, relatively small wavelength channel spacing is selected. To avoid significant inter-channel crosstalk under this condition, in addition to using MUX and DEMUX with narrow passband, precise wavelength control mechanism of the light sources is also required. Therefore, the cost and complexity of the system are relatively high. To avoid these issues, coarse WDM can be used in the system, where the channel spacing is much larger. In such systems, low-cost wavelength filters with larger passband can be used and no wavelength control mechanism is required, at the cost of reduced number of wavelength channels and lower aggregate data rates.

4.2 FDMA-, CDMA- and TDMA-Based Multi-User Access in Indoor OWC Systems

As discussed in the previous section, over 10 Gb/s wireless communication can be achieved in indoor OWC systems, and with wavelength multiplexing, even higher speed can be achieved. However, previous discussions and demonstrations focus on providing high-speed connectivity to a single user, whilst in practical applications, such as in offices, homes and shopping centers, multiple users need to be served simultaneously. Therefore, in this section, we introduce the common multi-user access techniques in indoor OWC systems. We will focus on the conventional techniques, including FDMA, CDMA and TDMA in this section, and the time-slot code (TSC)-based multi-user access principle will be discussed in the next section.

In traditional RF communication systems, access to multiple users is generally provided using multiplexing principles. The frequency domain, the code domain and the time domain can be explored, leading to the FDMA, CDMA and TDMA techniques, respectively. Such principles can also be used in indoor OWC systems to enable multi-user access.

In the FDMA-based multi-user access technique, the available bandwidth is divided into multiple sub-bands, and users are allocated with non-overlapping different sub-bands for data transmissions. Therefore, each user has dedicated spectrum and multiple users can be served simultaneously. FDMA is widely used in mobile communication systems and networks, and the principle can be used in indoor OWC systems as well. The optical version of frequency-division-multiplexing is usually in the form of subcarrier multiplexing (SCM), as shown in Figure 4.3 [9]. In the SCM-based multi-user access system, each user is allocated an RF subcarrier and these RF subcarriers have non-overlapping spectrum (i.e. $f_1, f_2, \ldots, f_n$). The data for each user is first modulated onto each RF subcarrier, and all data-carrying RF subcarriers are summed up. The spectrum of combined frequency-division-multiplexed RF signal is shown in the figure, and it is further modulated onto the optical carrier with the frequency of f_c. The optical signal contains a number of RF subcarriers, which carry dedicated data for multiple users, and it is then launched to the OWC link through a free-space interface.

After free-space propagation, the optical signal beam covers multiple users to provide simultaneous wireless access. At each OWC receiver, the light is captured by a receiver free-space interface, and it then passes through

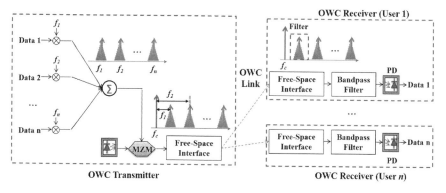

FIGURE 4.3
Principle of indoor OWC system with SCM-based multi-user access.

an optical bandpass filter. Each user is allocated with a dedicated optical bandpass filter, which only allows the RF subcarrier allocated to the user to pass and rejects all other RF subcarriers. The filtered signal is equivalent to a baseband signal, and it is then converted to the electrical domain with a PD for further processing and decision. Due to the baseband detection, the PD does not need to be ultra-broadband. Through this way, multiple users can be served by the OWC system simultaneously, and users share the available bandwidth in the frequency domain.

The SCM-based multi-access principle has been experimentally demonstrated. We consider the case of providing OWC to two users simultaneously for simplicity. In the SCM-based multi-access system, the first user can either use the baseband directly or be allocated an RF subcarrier f_1, and we select to use the baseband here for demonstration purposes. As shown in Figure 4.4, the two subcarrier modulated data streams are combined and then used to modulate the optical carrier. The optical carrier wavelength is 1551.72 nm, the RF subcarrier frequency for the second user is 10 GHz and the data rate for each user is 2.5 Gb/s with OOK format. Both RF spectrum and optical spectrum after modulation are shown in the figure, and it is clear that the power of subcarrier f_1 is about 7.8 dB lower than that of the baseband. The SCM-based optical signal then propagates in the free space via the free-space interface.

After OWC transmission, the optical signal is collected by the free-space interface at the receiver side. Then an optical bandpass filter is used to select the allocated RF subcarrier. Here the optical bandpass filter centered at 1551.72 nm is used for the first user, and the optical bandpass filter centered at subcarrier f_1 (i.e. 1551.72 nm + 10 GHz) is used for the second user. In order to select the dedicated subcarrier, the bandwidth of optical bandpass filter needs to be narrow. A multiple Fabry-Perot interferometer-based optical bandpass

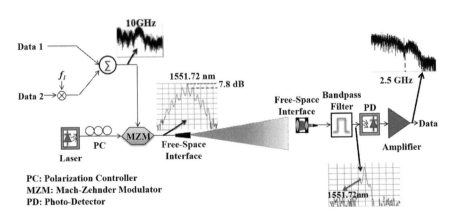

FIGURE 4.4
Demonstration of indoor OWC system with SCM-based multi-user access.

filter is used here and the bandwidth is about 6 GHz. The optical signal after bandpass filter for the second user is shown in Figure 4.4, and it is clear that the subcarrier dedicated for the user is filtered out and the signal for the other use is suppressed. Therefore, the baseband detection technique can be used to recover the data, which means a low-speed PD can be employed to reduce the cost at the user side. The RF spectrum of the signal after the PD is also shown in the figure. The converted signal is then amplified and detected.

The measured BER performance of the two users are shown in Figure 4.5. Similar with previous demonstrations, we characterize the BER performance as a function of the distance from beam coverage center. It is clear that the BER performances of the signals at baseband and at 10 GHz subcarrier band are different, where the baseband signal has better BER when the distance from beam center is the same. This agrees well with the optical spectrum shown in Figure 4.4, where the optical power at baseband is higher than that at the 10 GHz subcarrier band. Therefore, when the distance from beam center is the same, the baseband transmitted signal has higher power level, which results in lower BER. More importantly, using the SCM scheme, multiple users (two users here) can be served by the indoor OWC system simultaneously, providing the required multi-user access function.

In the SCM-based indoor OWC systems, the power spectrum needs to be evenly distributed across both baseband and subcarrier bands to use the available optical power efficiently and to provide fair service connections to multiple users. This can be realized by adding an optical filter that introduces higher loss for the baseband signal at the transmitter side after the modulator. One example is using the fiber Bragg grating (FBG) as shown in Figure 4.6. The FBG is a fiber-based component with periodic structure (i.e.

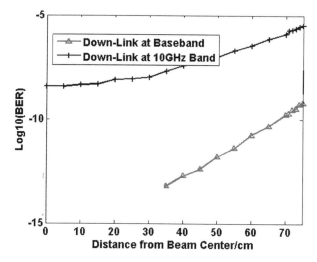

FIGURE 4.5
BER results of the indoor OWC system with SCM for two users.

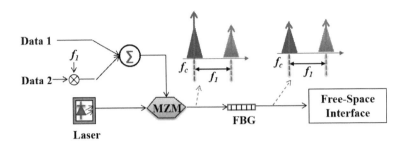

FIGURE 4.6
SCM-based indoor OWC system with FBG for spectrum flattening.

periodic change of refractive index), and it is based on signal diffractions that reflect the optical signal with the wavelength λ_f satisfying the Bragg condition, which can be expressed as:

$$\lambda_f = 2\Lambda n_{eff} \tag{4.1}$$

where Λ is the periodicity of the FBG and n_{eff} is the effective refractive index. As shown in Figure 4.6, with appropriately selected FBG, the optical spectrum of the SCM signal can be flattened, enabling evenly distributed power for all signals. In the demonstration discussed above, the power difference between the baseband and the subcarrier band can be reduced from about 7.8 dB to about 1.2 dB using the FBG. Therefore, the BER performance difference of the two users served by these two bands can be significantly reduced.

More subcarriers can be used in the SCM-based indoor OWC system to provide multi-access to a larger number of users. All users can transmit signals simultaneously without interference. However, the use of SCM for multi-access has three major limitations:

- First, to reduce the possible crosstalk between adjacent subcarriers, optical bandpass filters with narrow bandwidth are required at the receiver side to select the dedicated subcarriers, and the filters also need to have sharp roll-offs. Such filters are generally challenging to realize in the optical domain due to the high central frequency.
- Second, when the number of users in the system is large, the number of RF subcarriers is large as well. This results in a combined signal that occupies a wide bandwidth. To modulate such signal onto the optical carrier for free-space transmission, the optical modulator needs to have an ultra-broad operation bandwidth.
- Third, the discussion above uses the static subcarrier allocation to users. In practical scenarios, dynamic subcarrier allocation is required for efficient utilization of system resources. In this case, the

bandpass filter at the user side needs to be tunable. Due to the limited spacing between RF subcarriers, the tuning accuracy requirement of optical filters are very high, which further increases the filter design difficulty.

Due to the limitations of the FDMA principle (i.e. SCM in optical systems), other multi-access principles have also been proposed for indoor OWC system, such as the optical CDMA scheme [10]. In the optical CDMA scheme, each user is allocated a dedicated code sequence. Typically, the code occupies a sequence period T, which is divided into L time slots (also called chips) with a duration of T_C (i.e. $T = LT_C$). During the code sequence with code length L, only w positions are occupied by pulses (normally $w \ll L$), and the parameter w is called the weight of the code. When a data "1" needs to be transmitted, the user sends out the dedicated code sequence, and when the data is "0", the user does not send out anything.

The general principle of using OCDMA for the multi-user access in indoor OWC systems is shown in Figure 4.7. Dedicated code sequences are allocated to different users, and coded data streams are transmitted via the OWC link simultaneously. After free-space propagation, the optical signal is captured by each user, where the dedicated CDMA code is applied after photo-detection. Signals from other users can be suppressed after the CDMA decoding, and the decoded data is then further processed. Through this way, multiple users can be served simultaneously in the indoor OWC system.

In OCDMA-based indoor OWC systems, the design and selection of code sequences is critical. A number of code sequences have been proposed and demonstrated, such as the orthogonal optical codes, which have good correlation properties [14]. However, the implementation for a large number of

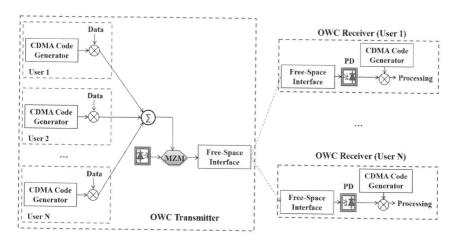

FIGURE 4.7
CDMA-based indoor OWC system.

users is challenging. Random optical codes (ROCs) have also been proposed to accommodate more users, whilst the correlation properties are relatively limited [10,15].

In addition to the FDMA- and CDMA-based multi-user access schemes, the TDMA principle can also be used in indoor OWC systems [11]. Different from other multi-user access schemes where all users can be served simultaneously, each user is served at different time instances in the TDMA-based system. The general principle of TDMA-based indoor OWC system is shown in Figure 4.8. N time slots are defined, and each user is assigned with one dedicated time slot. Each user is only allowed to transmit data in the dedicated time slot, and data streams from all users are combined. The combined signal is modulated to the optical carrier, which is launched to the OWC link via the free-space interface. At the user side, each user captures the signal light and converts it to the electrical domain with a PD. Then each user extracts the data from the dedicated time slot for further processing and discards data in all other time slots.

In TDMA-based indoor OWC systems, both static and dynamic time-slot allocation algorithms can be used. Guard band between adjacent time slots is normally required to reduce the possible crosstalk. The TDMA-based multi-user access principle is simple and straightforward. However, normally stringent timing and synchronization are required.

The FDMA-, CDMA- and TDMA-based principles described in this section are widely used in RF systems and networks, and these principles are adapted for indoor OWC systems. However, the use of traditional multi-user access schemes has several limitations, such as the high bandwidth requirements of optoelectronic device (i.e. modulator) in FDMA systems, the need for long-length code and complex processing techniques in CDMA systems, and the strict timing requirement in TDMA systems. To solve these issues, the TSC scheme can be utilized, and we will discuss TSC in the next section [12,13].

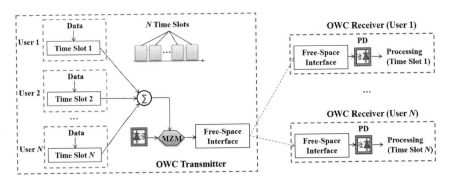

FIGURE 4.8
TDMA-based indoor OWC system.

4.3 TSC-Based Multi-User Access in Indoor OWC Systems

4.3.1 Time-Slot-Coding-Based Multi-User Access in Indoor OWC Systems

In order to overcome limitations of the conventional multi-user access techniques discussed in the previous section, in this section we introduce the TSC scheme to support multiple users simultaneously in high-speed indoor OWC systems. Similar with TDMA, the TSC scheme realizes the multi-access function in the time domain. However, in the traditional TDMA scheme, time synchronization issues are normally challenging [16] and it normally requires additional guard intervals or a scheduling framework to minimize the issue of time synchronization [17,18]. On the other hand, the TSC scheme overcomes these issues via a technique that uses pre-assigned code (referred to as time-slot code). As a result, in the TSC scheme, each user has a dedicated code for transmitting data, instead of using a timing window as in the conventional TDMA implementations. Therefore, the TSC scheme does not require scheduling framework or guard intervals to allow the transmission of data streams from multiple users over different time slots.

The basic principle of the TSC-based multi-user access in indoor OWC systems is shown in Figure 4.9. Data streams for multiple users are first baseband modulated. Assume the original symbol period of each user is T_s and all users have the same symbol period. There are a total number of n users to be connected by the OWC link in the system. The time-slot codes applied to multiple users are a set of n-bit codes with only unipolar values (i.e. 0 or 1). The code allocated to each user has a unique location for the value "1", and time-slot code can be expressed as an $n \times n$ matrix:

$$C_n = \begin{bmatrix} 1 & 0 & \cdots & 0 \\ 0 & 1 & \cdots & 0 \\ \vdots & \vdots & \ddots & \vdots \\ 0 & 0 & \cdots & 1 \end{bmatrix} \qquad (4.2)$$

where each row of the matrix represents the time-slot code allocated to an individual user.

In the system, the ith user is assigned with the ith row of the matrix shown by Eq. (4.2), where the ith code bit is the location for the value of "1". Each user's symbol sequence v_i is then multiplied with the dedicated time-slot code. The original symbol data sequences and the coded sequences are illustrated in Figure 4.10. The coded data sequences from all the users are then added together to form the transmission data. With these codes, each user is ideally capable of occupying one non-overlapping time slot to avoid

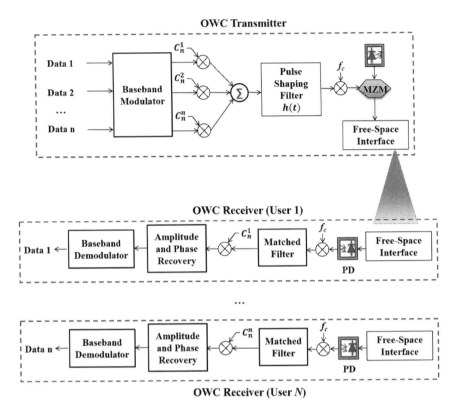

FIGURE 4.9
Principle of the TSC-based indoor OWC system for multi-user access.

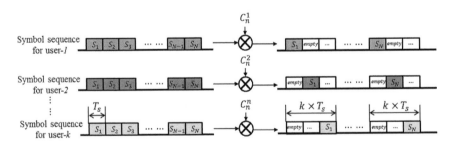

FIGURE 4.10
Original symbol sequence and coded symbol sequence after TSC.

multi-user interference (MUI), when all users are served at the same time. The combined symbol sequence v_c then passes through a pulse shaping filter $h(t)$ and the raised-cosine pulse shape is selected here. After the filter, the output sequence s_m can be expressed as:

$$s_m = \sum_{j=-\infty}^{\infty} v_c h(m - jT_s) \tag{4.3}$$

We assume the baseband modulation format in the system is quadrature amplitude modulation (QAM). Due to the lack of IQ mixer and IQ modulator here, the baseband QAM signal is upconverted to a carrier frequency f_c and the carrier frequency is selected at 1 GHz. The QAM signal then modulates the laser source via an external modulator, and the modulated optical signal carrying data streams for multiple users is launched into the OWC link. After free-space propagation, the receiver optics captures the signal light and the PD is employed at the user terminal to detect the optical signal. The detected signal is first down converted to the baseband by multiplying it with the local oscillator with the frequency f_c, and the baseband signal passes through the matched filter. The user-specific time-slot code is then applied to each user to obtain its original symbol sequence. Through this way, multi-user access can be achieved using the TSC scheme.

The proposed TSC-based multi-user access in indoor OWC systems can be experimentally demonstrated. Here we consider the case that the data rate requirement of all users is the same. The baseband modulation format is 4-QAM, the optical carrier wavelength is 1548 nm and the 3-dB bandwidth of PD is 2 GHz. Similar with previous demonstrations, we characterize the BER performance of all users as a function of the distance from beam coverage center. When there are a total number of 5 users and 8 users, the BER results of all users are shown in Figure 4.11a and b, respectively. The aggregate data rate of all users is kept constant in both scenarios, which is selected at 2.5 Gb/s. As shown in the figure, regardless of the total number of users connected simultaneously, there is negligible penalty amongst the users with all users exhibiting similar performance. This is mainly due to the fact that all users are treated fairly, as specified in the proposed TSC scheme (i.e. there is only one value "1" in the code allocated to each user). In addition, comparing the average BER obtained with 5 users and 8 users, it can be seen that the

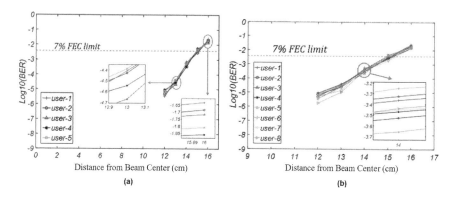

FIGURE 4.11
Measured the BER performance of all users with the TSC scheme. The total number of users is: (a) 5 users and (b) 8 users.

BER performance does not change with the number of users. Therefore, even more users can be served with the TSC scheme simultaneously, providing a promising multi-user access solution in indoor OWC systems.

4.3.2 Adaptive Loading-Based Time-Slot Coding for Multi-User Access

In the TSC-based multi-user access scheme discussed in the previous section, all users, regardless of the location inside the beam coverage area, are connected by the indoor OWC system with the same data rate. Whilst such configuration treats all users fairly, the optical power available is not used efficiently. This is because of the optical intensity distribution across the beam coverage area, where usually the intensity is higher at the beam center and becomes lower towards the beam edges. Therefore, for users located close to the beam center, a large signal power is available, and the data rate supported can be higher.

To accommodate this property, here we propose the adaptive loading-based TSC scheme for indoor OWC systems [13]. Different data rates are achieved by using different modulation formats. Here, the transmitter assigns higher data rates with more spectral efficient advanced modulation format to users closer to the beam center and assigns lower data rates with lower-order modulation format to users further away from the beam center. To realize this adaptive loading, it is clear that the location information of users is required, and it can be provided by indoor localization technologies [19,20], as discussed in Section 2.3.

We still consider the TSC demonstration example discussed in the previous section and we still assume that the total number of users to be connected is 5. Here we further assume that user-2, user-4 and user-5 are located closer to the beam center, whilst user-1 and user-3 are located further away towards the beam edges, as shown in Figure 4.12. Applying the adaptive loading principle, we assign 16-QAM modulated format to users 2, 4 and 5 and assign 4-QAM modulation format to users 1 and 3.

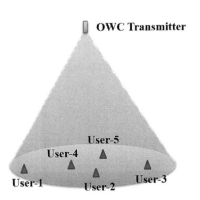

FIGURE 4.12
Location of multi-users considered for adaptive loading.

In the adaptive loading scheme where 4-QAM and 16-QAM modulation formats are used, we considered two possible transmission strategies. The first strategy (S_1) is so selected that the minimum Euclidean distance of 4-QAM (d_1) is normalized to three times of that in 16-QAM (d_2). Therefore, $d_1 = 3d_2$ and the transmission power available can be fully used for both 4-QAM and 16-QAM formats. In the second strategy (S_2), the standard 4-QAM constellation is used, where both modulation formats have the same minimum Euclidean distance (i.e. $d_1 = d_2$). The two strategies are shown in Figure 4.13.

The measured BER of all users with these two transmission strategies are shown in Figure 4.14. The symbol rate in the system is 1 GBaud/s. Therefore, the effective data rate for users 2, 4 and 5 is 4 Gb/s and it is 2 Gb/s for users 1 and 5. It can be observed that the performances of user-2, user-4 and user-5 served with 16-QAM do not vary noticeably from strategy S_1 to S_2. This is because as shown by Figure 4.13, the 16-QAM symbol constellation remains the same in both strategies. On the other hand, the performance of user-1 and user-3 served by 4-QAM has different BER performance under the two strategies, where S_1 provides better performance. This is because that the 4-QAM modulation format in S_2 does not take full advantage of the transmission power due to the smaller minimum Euclidean distance. More importantly, as shown by the results, all five users served by different modulation formats according to their locations can achieve error-free operation (with FEC). The total effective data rate of the adaptive loading scheme is 3.2 Gb/s (4 Gb/s for three users and 2 Gb/s for two users),

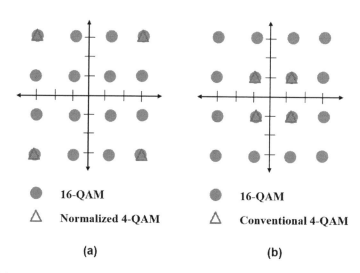

| 16-QAM | 16-QAM |
| Normalized 4-QAM | Conventional 4-QAM |

(a)   (b)

FIGURE 4.13
Constellations of the 4-QAM and 16-QAM symbols. (a) Strategy 1 (i.e. $d_1 = 3d_2$) and (b) strategy 2 (i.e. $d_1 = d_2$).

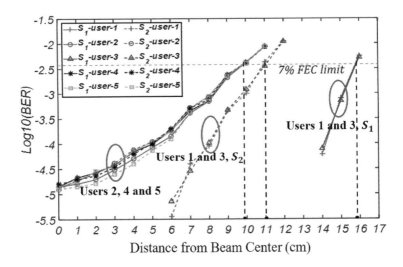

FIGURE 4.14
The TSC BER performance with adaptive loading function: user-1 (4-QAM), user-2 (16-QAM), user-3 (4-QAM), user-4 (16-QAM), user-5 (16-QAM).

increased from 2 Gb/s in the case of serving all users with uniform 4-QAM format. This improvement is achieved by providing higher-order modulation to users with larger optical signal power.

From the analysis and demonstration above, it can be seen that the adaptive loading-based TSC scheme can use the signal power more efficiently, to achieve higher system throughput when providing multi-user access in indoor OWC systems. The principle discussed above can be applied to more users using more versatile modulation formats.

4.3.3 TSC Code Misalignment Tolerance

In the previous sections, we have introduced and demonstrated the TSC scheme with ideal code set as shown by Eq. (4.2) for the multi-user access in indoor OWC systems. As discussed in the principle of TSC scheme and shown in Figure 4.15a, each active user (a total of four users shown in the figure) can employ dedicated ideal time-slot code to avoid the interference from other users by occupying non-overlapping time slots.

However, the hardware implementation procedures during the code generation process are non-ideal. Therefore, the time-slot codes assigned to users may have misalignment in the time domain, and the timing issue potentially leads to partial overlap within the time-slot code set. This code misalignment issue is shown in Figure 4.15b, where there is significant overlap between user-1 and user-2 and between user-3 and user-4. The code misalignment normally results in multi-user interference and degrades the OWC system performance. In this section, we study the effect of code overlap on the BER

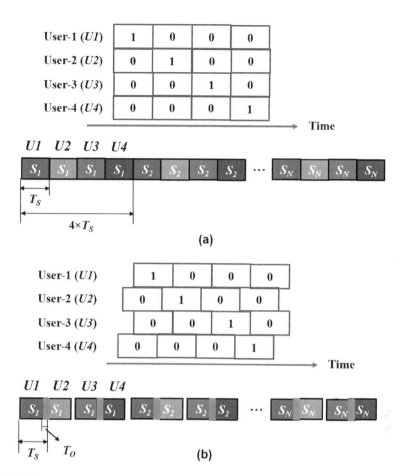

FIGURE 4.15
Illustration of the TSC scheme: (a) with ideal code and (b) with misaligned code.

performance in indoor OWC systems based on a general square QAM modulation format. The analytical process here can be easily transferred to other modulation formats. The advantage of using the TSC scheme compared to the conventional TDMA to provide multi-user access is also analyzed.

The combined data sequence in the TSC scheme with timing misalignment issue is shown in Figure 4.15b. Four users are considered here. Due to the misalignment, significant code overlap between users is clear from the figure, where the gray shaded area is the overlapping between neighboring symbols from different users. T_O here denotes the total time duration of the overlapping from adjacent users. Accordingly, we can define the overlapping ratio R_O as the ratio of the total overlapping period and the original symbol period:

$$R_O = \frac{T_O}{T_S}, \qquad R_O \in [0,1] \tag{4.4}$$

We can then define the code misalignment tolerance as the maximum value of R_O when error-free performance can still be achieved. The received signal suffering from code overlapping after frequency down-conversion in the system can be described as:

$$r_m = \sum_{j=-\infty}^{\infty} S \cdot \left(R_O^{j-1} v_c^{j-1} + v_c^j + R_O^{j+1} v_c^{j+1} \right) h \left(m - jT_S \right) + n_m \tag{4.5}$$

where S is the signal amplitude, R_O^{j-1} represents the overlapping ratio caused by the preceding symbol v_c^{j-1}, R_O^{j+1} is the overlapping ratio introduced by the following symbol v_c^{j+1}, $h(t)$ is the response of the pulse shaping filter and n_m is the noise introduced by the free-space transmission and detection. Using this equation, the pth received symbol of the kth user can be expressed as:

$$y_k[p] = S \cdot \left(R_O^{j-1} v_{k-1}^p + v_k^p + R_O^{j+1} v_{k+1}^p \right) + n' \tag{4.6}$$

where v_k^p represents the original pth symbol of the kth user, v_{k-1}^p denotes the original pth symbol of the $(k-1)$th user, v_{k+1}^p denotes the original pth symbol of the $(k+1)$th user and n' is the noise of symbol. The noise applied to the overlapped symbol sequence after detection is assumed to follow a Gaussian distribution. The received symbol as described in Eq. (4.6) then can be analyzed theoretically using the system model established in Chapter 2.

For simplicity, here we consider a simple scenario where two users are served in the indoor OWC system with the TSC scheme. Both users use the 4-QAM modulation format. Since the four I-Q quadrants in the signal constellation diagram are symmetrical, we only show the first quadrant for clear explanation in Figure 4.16. The detailed analysis steps are described as follows:

Step 1: After the code generation process, the overlapping ratio R_O between the two users is fixed. The symbol (1, 1) is possibly overlapped by all four symbol patterns in the 4-QAM modulation format. As shown in Figure 4.16a, the gray star represents the original symbol and the black icons represent the overlapping symbols without considering the system noise. For example, the original symbol (1, 1) is overlapped by another symbol (1, –1) with the overlapping ratio R_O and the resultant overlapping symbol can be then expressed as $(1 + 1 \cdot R_O, 1 - 1 \cdot R_O)$.

Step 2: When the system noise is considered, the noisy version of original symbol and overlapping symbols after transmission and detection are shown in Figure 4.16b. The noise around each ideal overlapping symbol follows two-dimensional (2D) Gaussian distribution $f(x, y)$, where variables x and y are independent and have the same Gaussian distribution with zero mean and the standard deviation σ.

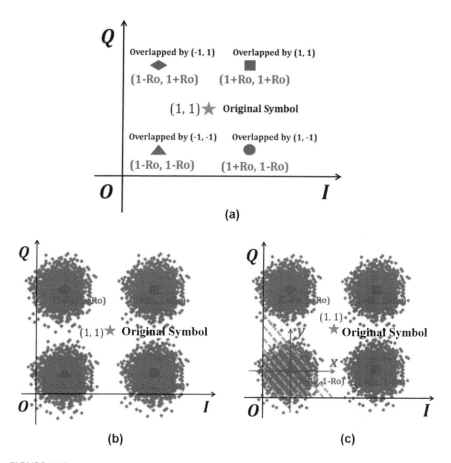

FIGURE 4.16
Illustration of the code misalignment in the TSC scheme. (a) Illustration of possible overlapping for symbol (1, 1) without noise; (b) illustration of possible overlapping for symbol (1, 1) with noise; and (c) illustration of the analysis of symbol overlapping.

Step 3: Take one overlapping symbol, $(1-1\cdot R_O, 1-1\cdot R_O)$, as an example. As shown in Figure 4.16c, we set the location of the ideal overlapping symbol as the origin of a new $x-y$ coordinate. The rate that received symbols are able to be correctly demodulated as symbol (1, 1), which is denoted as R_1, can be calculated by taking an integration of both x and y over the decision region of the reference symbol (1, 1) as shown in the dash-line region [21]:

$$R_1 = \int_{-(1-R_O)}^{\infty} \int_{-(1-R_O)}^{\infty} f(x,y)\,dxdy \qquad (4.7)$$

Step 4: Apply the same procedures as described above to the other three overlapping symbols as well as to the rest of quadrants, the symbol-error-rate (SER) with respect to the overlapping ratio R_O for the 4-QAM modulation format can be calculated as:

$$SER_{4-QAM}\left(R_O,\sigma\right)=1-\frac{1}{4}\sum_{I,Q}\int_{-(1+I\cdot R_O)}^{\infty}\int_{-(1+Q\cdot R_O)}^{\infty}f\left(x,y\right)dxdy \qquad (4.8)$$

where *I* and *Q* equal to either 1 or –1 in the 4-QAM modulation format.

Using the steps described above, the SER of a general 2^{2l}–QAM can be derived. We skip the detailed steps here and the result is given in Eq. (4.9):

$$SER\left(l,R_O,\sigma\right)=1-\frac{4}{\left(2^{2l}\right)^2}\left[\begin{array}{l}\sum_{c=1,d=1}\sum_{I_2,Q_2}\int_{Q_{1,d}-1-y_{N,d}}^{Q_{1,d}+1-y_{N,d}}\int_{I_{1,c}-1-x_{N,c}}^{I_{1,c}+1-x_{N,c}}f\left(x,y\right)dxdy\\[4pt]+2\sum_{c=1,d=2}\sum_{I_2,Q_2}\int_{Q_{1,d}-1-y_{N,d}}^{\infty}\int_{I_{1,c}-1-x_{N,c}}^{I_{1,c}+1-x_{N,c}}f\left(x,y\right)dxdy\\[4pt]+\sum_{c=2,d=2}\sum_{I_2,Q_2}\int_{Q_{1,d}-1-y_{N,d}}^{\infty}\int_{I_{1,c}-1-x_{N,c}}^{\infty}f\left(x,y\right)dxdy\end{array}\right]$$

$$=1-\frac{4}{\left(2^{2l}\right)^2}\left[\begin{array}{l}\sum_{c=1,d=1}\sum_{I_2,Q_2}\begin{array}{l}\left(F\left(Q_{1,d}+1-y_{N,d},\sigma\right)-F\left(Q_{1,d}-1-y_{N,d},\sigma\right)\right)\cdot\\\left(F\left(I_{1,c}+1-x_{N,c},\sigma\right)-F\left(I_{1,c}-1-x_{N,c},\sigma\right)\right)\end{array}\\[6pt]+2\sum_{c=1,d=2}\sum_{I_2,Q_2}\begin{array}{l}F\left(-\left(Q_{1,d}-1-y_{N,d}\right),\sigma\right)\cdot\\\left(F\left(I_{1,c}+1-x_{N,c},\sigma\right)-F\left(I_{1,c}-1-x_{N,c},\sigma\right)\right)\end{array}\\[6pt]+\sum_{c=2,d=2}\sum_{I_2,Q_2}F\left(-\left(Q_{1,d}-1-y_{N,d}\right),\sigma\right)\cdot F\left(-\left(I_{1,c}-1-x_{N,c}\right),\sigma\right)\end{array}\right]$$

$$(4.9)$$

where $f(x,y)$ is the probability density function of the received signal position and $F(\omega,\sigma)$ can be expressed as:

$$F\left(\omega,\sigma\right)=\int_{-\infty}^{\infty}\frac{1}{\sigma\sqrt{2\pi}}\exp\left(-\frac{\omega^2}{2\sigma^2}\right)d\omega \qquad (4.10)$$

$I_{1,c}, Q_{1,d}, I_2$ and Q_2 in Eq. (4.9) can be expressed as follows for a general square 2^{2l}–QAM modulation format (where *l* is a natural number, i.e. $l=1,2,3,...$):

$$I_{1,1}=2a-1-2^l,\quad 2^{l-1}<a<2^l \qquad (4.11)$$

$$Q_{1,1} = 2b - 1 - 2^l, \quad 2^{l-1} < b < 2^l \tag{4.12}$$

$$I_{1,2} = 2^l - 1 \tag{4.13}$$

$$Q_{1,2} = 2^l - 1 \tag{4.14}$$

$$I_2 = 2p - 1 - 2^l, \quad 1 \le p \le 2^l \tag{4.15}$$

$$Q_2 = 2q - 1 - 2^l, \quad 1 \le q \le 2^l \tag{4.16}$$

where a, b, p and q are natural numbers. The notation of $x_{N,c}$ and $y_{N,d}$ for the TSC scheme with code overlapping can be derived as:

$$x_{N,c} = I_{1,c} + R_O I_2 \tag{4.17}$$

$$y_{N,d} = Q_{1,d} + R_O Q_2 \tag{4.18}$$

where $c, d = 1$ or 2, $(I_{1,c}, Q_{1,d})$ is the user's original symbol and (I_2, Q_2) is the symbol that partially overlaps with the original symbol due to code misalignment.

To analyze the advantage of the TSC scheme compared to the conventional TDMA technique discussed in Section 4.2, we can further derive the SER of TDMA with asynchronized time window, which is resulted when obtaining the symbol data. For fair comparison, no guard interval is included in either TSC or TDMA implementations. The TDMA with time window overlapping is shown in Figure 4.17, where k users each with N symbols are considered. The overlapping ratio R_O is the same as that defined in Eq. (4.4). Following the same procedure as describe before, $x_{N,c}$ and $y_{N,d}$ of TDMA with asynchronized time window can be expressed as:

$$x_{N,c} = (1 - R_O) I_{1,c} + R_O I_2 \tag{4.19}$$

$$y_{N,d} = (1 - R_O) Q_{1,d} + R_O Q_2 \tag{4.20}$$

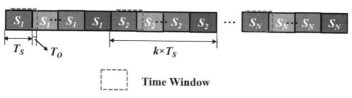

Time Window

FIGURE 4.17
TDMA with asynchronized time window.

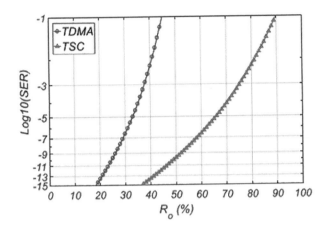

FIGURE 4.18

Analytical SER performance of TSC and TDMA schemes with respect to the overlapping ratio R_O.

By substituting Eq. (4.17) – Eq. (4.20) to Eq. (4.9), the SER of both TSC and conventional TDMA schemes with respect to the overlapping ratio R_O can be calculated. Assume the noise variance σ is fixed at 0.08 for both multi-user access schemes, we plot the analytical SER results in Figure 4.18. The modulation format in the analysis is 4-QAM.

From the analytical results, it can be seen that the TSC scheme is much more tolerant to the overlapping between adjacent symbols than the conventional TDMA. This advantage is mainly enabled by the difference in the received symbol sequence with imperfect timing. In the TSC scheme, each user is able to obtain the entire combined symbol sequence. In contrast, the user in the TDMA scheme only obtains partial combined symbol sequence after applying the dedicated time window. Therefore, the interference from adjacent symbols can be better suppressed in the TSC-based system.

The study presented here is based on the 4-QAM modulation format, and similar results can be obtained for other modulation formats. In general, the TSC scheme is much more tolerant against the imperfect timing when providing multi-user access in indoor OWC systems, which is highly likely to exist due to moving users and imperfect hardware. Therefore, TSC provides a promising solution in practical indoor OWC systems when multi-user access is required.

4.4 Conclusions

In this chapter, we have described the wavelength multiplexing and the multi-user access technologies in near-infrared high-speed indoor OWC systems. The wavelength multiplexing has been utilized to enable the parallel transmission of multiple data streams, which can increase the aggregate communication data rate and overall capacity of the system. The general technology, system structure, operation principle and a demonstration example have been presented, and the trade-off between the channel spacing and the achievable data rate has been discussed in Section 4.1.

In addition to providing high-speed wireless communication to a single user, practically multiple users need to be served simultaneously in indoor OWC systems. Therefore, we have reviewed multi-user access techniques in this chapter. The conventional multi-user access principles widely used in RF systems, including FDMA, CDMA and TDMA, have also been applied in indoor OWC systems. Therefore, the basic principles and demonstration examples have been discussed in Section 4.2. The major limitations of these conventional schemes have also been analyzed.

To accommodate the characteristics of indoor OWC systems, the TSC-based multi-user access principle has been described in Section 4.3. In addition to the basic principles and demonstration examples, we have also discussed the advanced adaptive loading technique in the TSC-based multi-user system, where users with different channel conditions are served with data rates adaptively. It has been shown that the adaptive loading can be achieved using adaptive modulation formats, and it can increase the effective system data rate significantly. Furthermore, the code misalignment issue due to moving users and imperfect hardware in TSC-based indoor OWC systems has also been analyzed theoretically. It has been shown that the TSC scheme is tolerant against imperfect timing, and it can achieve much better performance compared to the conventional TDMA scheme.

References

1. K. Wang, A. Nirmalathas, C. Lim, and E. Skafidas, High-speed duplex optical wireless communication system for indoor personal area networks. *Optics Express*, 2010. **18**(24): pp. 52199–25216.
2. K. Wang, A. Nirmalathas, C. Lim, and E. Skafidas, High-speed optical wireless communication system for indoor applications. *IEEE Photonics Technology Letters*, 2011. **23**(8): pp. 519–521.
3. F. Gomez-Agis, S.P. Van de Heide, C.M. Okonkwo, E. Tangdiongga, and A.M.J. Koonen, 112 Gbit/s transmission in a 2D beam steering AWG-based optical wireless communication system, in *European Conference on Optical Communications (ECOC)*, Goteborg, Sweden, 2017.
4. K. Wang, A. Nirmalathas, C. Lim, K. Alameh, and E. Skafidas, Full-duplex gigabit indoor optical wireless communication system with CAP modulation. *IEEE Photonics Technology Letters*, 2016. **28**(7): pp. 790–793.
5. N. Huang, J.-B. Wang, J. Wang, C. Pan, H. Wang, and M. Chen, Receiver design for PAM-DMT in indoor optical wireless links. *IEEE Photonics Technology Letters*, 2015. **27**(2): pp. 161–164.

6. A.H. Azhar, T.-A.Tarn, and D. O'Brien, A gigabit/s indoor wireless transmission using MIMO-OFDM visible-light communications. *IEEE Photonics Technology Letters*, 2013. **25**(2): pp. 171–174.

7. K. Wang, A. Nirmalathas, C. Lim, and E. Skafidas, High speed 4× 12.5 Gbps WDM optical wireless communication systems for indoor applications, in *Optical Fiber Communication Conference*, OSA Publishing, 2011.

8. Z. Cao, L. Shen, Y. Jiao, X. Zhao, and T. Koonen, 200 Gbps OOK transmission over an indoor optical wireless link enabled by an integrated cascaded aperture optical receiver, in *Optical Fiber Communications Conference (OFC)*, OSA Publishing, Los Angeles, CA, 2017.

9. K. Wang, A. Nirmalathas, C. Lim, and E. Skafidas, Indoor gigabit full-duplex optical wireless communication system with SCM based multiple user access. in *International Topical Meeting on Microwave Photonics (MWP)*, IEEE, Singapore, 2011.

10. F.A. Delgado-Rajó, O. González, J.A. Martin-Gonzalez, M.F. Guerra-Medina, and F.J. López-Hernández, Cyclic code-shift extension keying for multi-user optical wireless communications. *Electronics Letters*, 2015. **51**(11): pp. 847–849.

11. T. Song, K. Wang, J. Ma, and A. Nirmalathas, Experimental demonstration of optical wireless personal area communication system supporting multiple users, in *Optical Fiber Communication Conference (OFC)*, OSA Publishing, Anaheim, CA, 2016.

12. T. Liang, K. Wang, C. Lim, E. Wong, T. Song, and A. Nirmalathas, Time-slot coding scheme for multiple access in indoor optical wireless communications. *Optics Letters*, 2016. **41**(22): pp. 5166–5169.

13. T. Liang, K. Wang, C. Lim, E. Wong, T. Song, and A. Nirmalathas, Time-slot coding scheme with adaptive loading function for multiple access in indoor optical wireless communications. *Journal of Lightwave Technology*, 2017. **35**(18): pp. 4079–4086.

14. J.A. Salehi, Emerging OCDMA communication systems and data networks. *Journal of Optical Networking*, 2007. **6**(9): pp. 1138–1178.

15. F.J. Lopez-Hernandez, J.A. Martin-Gonzalez, and E. Poves, Random optical codes used in optical networks. *IET Communications*, 2009. **3**(8): pp. 1392–1401.

16. J.C. Chuang, The effects of port synchronization on TDMA personal communications for tow duplexing methods, in *Vehicular Technology Conference*, IEEE, Denver, CO, 1992.

17. P.P. Nuspl, K. E. Brown, W. Steenaart, and B. Ghicopoulos, Synchronization methods for TDMA. *Proceedings of IEEE*, 1977. **65**(3): pp. 434–444.

18. B. Yener and F. Sivrikaya, Time synchronization in sensor network: A survey. *IEEE Networking*, 2004. **18**(4): pp. 45–50.

19. S.-H. Yang, H.-S. Kim, Y.-H. Son, and S.-K. Han, Three-dimensional visible light indoor localization using AOA and RSS with multiple optical receivers. *Journal of Lightwave Technology*, 2014. **32**(14): pp. 2480–2485.

20. K. Wang, A. Nirmalathas, C. Lim, K. Alameh, and E. Skafidas, Optical wireless-based indoor localization system employing a single-channel imaging receiver. *Journal of Lightwave Technology*, 2016. **34**(4): pp. 1141–1149.

21. R.A. Shafik, M.S. Rahman, and A.R. Islam. On the extended relationships among EVM, BER and SNR as performance metrics, in *International Conference on Electrical and Computer Engineering (ICECE)*, IEEE, 2006.

5

Photonic Integrations of Near-Infrared Indoor Optical Wireless Communications

In previous chapters, we have established the general principles and models of near-infrared indoor OWC systems, including both direct LOS link- and diffusive link-based approaches. We have also introduced more advanced techniques for indoor OWC systems, such as the spatial diversity principle to improve the system robustness and the wavelength multiplexing scheme to increase the overall communication data rate. The multi-user access principle and solutions to serve multiple users simultaneously have also been discussed.

However, in all previous chapters, the indoor OWC system discussed is based on the use of multiple discrete components and devices. For example, in the system shown in Figure 3.6, the laser, the modulator, the free-space interfaces at both transmitter and receiver sides and the PD are all separate devices and components. These devices and components are connected together with optical fibers or through free-space coupling in the system. Therefore, the OWC transceiver and system are usually bulky, sensitive to coupling errors, and expensive.

To solve these limitations, photonic integration technologies have been proposed and studied. Photonic integrations aim to use advanced semiconductor technologies to realize integrated components, functions, circuits and transceivers for optical systems. Generally, the resulted systems are compact, and the cost can be significantly reduced via mass production. The system performance can be improved through photonic integrations as well, mainly due to the better signal coupling and interfaces between different building blocks. In this chapter, we will briefly introduce the photonic integration platform for the integration of near-infrared indoor OWC systems.

This chapter is organized as follows: the basics of photonic integration technologies are introduced in Section 5.1; integrated key components and devices in indoor OWC systems are discussed in Sections 5.2 and 5.3, where passive devices and active devices are discussed, respectively; the photonic integrated beam steering device, which is unique for indoor OWC systems for the control of free-space propagation direction, is introduced in Section 5.4; the experimental demonstration of the high-speed indoor OWC system with photonic integrated circuit is then provided in Section 5.5, and finally, this chapter is summarized in Section 5.6.

5.1 Introduction of Photonic Integrations

Optical components nowadays are mainly discrete components fabricated and packaged independently. For example, optical modulators at the 1550 nm band is normally based on LiNbO$_3$, which has outstanding electro-optic effect, and each modulator is packaged with optical and electrical interfaces. At the receiver side, 1550 nm PDs are widely designed, fabricated and packaged based on the InGaAs platform, which has high response and high-speed properties. However, these two material platforms are incompatible with each other, and hence, the modulators and PDs cannot be fabricated and packaged together.

The use of discrete components in optical systems results in a number of limitations. First, different components need to be designed, optimized, fabricated and packaged separately, requiring significant engineering efforts and leading to high costs of systems. Second, separate components need to be connected together using optical fibers or via free-space interfaces, resulting in high signal coupling losses and hence, limiting the system performance. This also leads to bulky systems, and additional attention is required to ensure the coupling and system stability. Third, different components need to be controlled individually in optical systems, resulting in additional complexity and challenges, which usually limits the number of components in practical systems.

If we look back at the development history of electronic components and systems, we can easily see that electronic systems in early days are also based on discrete components, whilst integration technologies enable current advanced electronic systems. Nowadays, millions or even billions of transistors can be integrated onto a centimeter- or even millimeter-sized electronic integrated circuit on the same substrate (mostly based on silicon), and numerous functions can be realized simultaneously with high-speed. In addition, with integrated circuit technologies, powerful electronic integrated circuits are now available at low cost.

Inspired by electronic integrated circuits, the possibility and the development of photonic integrated circuits have attracted numerous attentions in the past decade [1]. Similar with electronic integrations, the photonic integration technology integrates multiple photonic functions or devices onto the same integrated circuit with the same substrate [2]. An example of photonic integrated circuit is shown in Figure 5.1. It is a WDM optical transceiver, where the light sources, optical modulators, WDM MUX, optical bandpass filter, WDM DEMUX and PDs are all integrated together.

The use of photonic integration technologies has many unique advantages. First, a large number of components can be designed and fabricated simultaneously using existing and advanced semiconductor fabrication processes, and hence, the cost of optical functions, transceivers and systems can

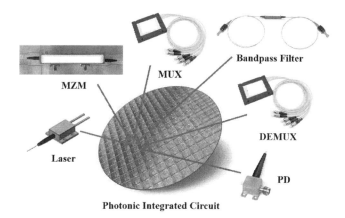

FIGURE 5.1
The basic concept of photonic integration technology, where multiple optical components are integrated onto the same circuit using the same substrate.

be significantly reduced. Second, all components are connected on the same integrated circuit with optical waveguides, where have low-loss with appropriate designs. Therefore, the stability of the integrated optical system is much better than that in the discrete components-based solution, and the signal coupling between components is improved. Third, all components integrated on the same photonic circuit can be controlled individually and at the same time, simultaneously. It facilitates the realization of more complicated functions and more accurate controls. Fourthly, the number of components that can be used in an optical transceiver and system is increased significantly using the photonic integration technology. Therefore, larger-scale optical systems can be realized with reduced complexity. In addition, with the photonic integration technology, the physical size, weight and power consumption of optical transceivers and systems are reduced substantially, enabling more application scenarios, such as in mobile and low-profile conditions.

A number of semiconductor materials platforms can be used to realize the photonic integrated circuits and systems. For example, indium phosphate (InP) is a promising integration platform for optical signal generation, switching and detection, especially for those operating in the 1300–1600 nm wavelength region [3]. Another promising photonic integration platform that has attracted lots of attention is Lithium Niobate on Insulator (LNOI), which has robust and low-loss transparent operation window ranging from the visible light to the mid-infrared wavelength region [4]. In addition, it has both strong electro-optical coefficient and high second-order optical nonlinearity.

In addition to the InP- and LNOI-based photonic integration platforms, the use of silicon (Si) platform has attracted even more interests from both research and industry communities. This is mainly due to the unique advantages of using Si as the photonic integration platform. Some key advantages include:

- First, existing and advanced electronic integrated circuits are designed and fabricated based on the Si CMOS process. Therefore, by using the same Si platform, photonic integration technologies can leverage with CMOS, where highly advanced fabrication tools and processes can be re-used, to save significant engineering efforts and infrastructure investments.

- Second, Si is a reasonably good material for photonic components and functions. For example, Si has a bandgap of 1.14 eV, and hence, it is transparent in the 1550 nm wavelength region, which is the most widely used band with mature technologies. The refractive index of Si is also high (over 3.4 in the 1550). With a large refractive index contrast, the optical signal can be tightly confined in the silicon structure, and compact size of integrated devices and transceivers can be achieved. In addition, although Si does not have good electro-optic effect for optical modulations, it has the plasma dispersion effect, which can be explored for modulation [5]. Other major advantages of Si for optical integrations include the high thermal conductivity and the high optical damage threshold.

- Third, due to the large refractive index contrast in Si photonic integrations, the optical signal is confined in a small region and the light intensity is high. Therefore, nonlinear optical effects can be activated with a low power threshold, such as Raman and Kerr effect. These effects can be explored for versatile optical functions, such as optical amplification and wavelength conversion.

In addition, with the silicon integration platform, both optical and electronic components and functions can be integrated on the same circuit (although with some challenges to be solved). Therefore, the strengths provided by both fields can be explored simultaneously, such as the combining the powerful signal processing capability provided by electronics with the broadband signal transmission property offered by optics. Due to these unique advantages, the silicon photonic integration technology has been widely considered as the solution for future optical transceivers and systems, especially for applications in data center interconnects and integrated sensors.

5.2 Silicon Photonic Integrated Passive Devices

Due to the advantages of the silicon photonic integration technology, it has attracted intensive interests and most basic optical devices and functions have been demonstrated. In optical transceivers and systems, devices can be divided into passive and active categories, depending on if external power is required for the operation. Major passive devices include basic signal waveguides, power splitters, optical filters, WDM MUX and DEMUX, and optical polarization controllers; and key active devices include optical sources, optical amplifiers, modulators, optical switches and PDs. In this section, we introduce key passive devices based on the silicon photonic integration platform. Active integrated devices will be discussed in the next section.

5.2.1 Silicon Waveguides

In silicon integrated photonic circuits, optical signals propagate in silicon waveguides, which provide the light confinement and transmission function. The silicon waveguide also works as the signal guiding block in other optical devices, such as optical filters and modulators. To enable efficient signal transmission and other optical functions, one key requirement of silicon waveguide is the low signal propagation loss. In addition, relatively sharp bends of waveguides are desirable to enable flexible circuit layout and compact circuit size.

In order to provide light confinement, refractive index contrast is a necessity. Therefore, the silicon photonic integration technology uses the silicon-on-insulator (SOI) platform, where a buried silicon oxide (BOX) layer is formed underneath the device silicon layer to provide the vertical confinement, as shown in Figure 5.2. The top cladding of the device silicon layer can be either air or SiO_2, and the use of air top cladding is shown in the figure.

Based on the SOI platform, two types of silicon waveguides are widely used to realize the key functions and to satisfy the requirements mentioned above (i.e. the strip waveguide and the rib waveguide), which are also shown in Figure 5.2. In the strip waveguide, which is also referred as wire waveguide, a rectangular waveguide core is formed directly above the BOX layer and

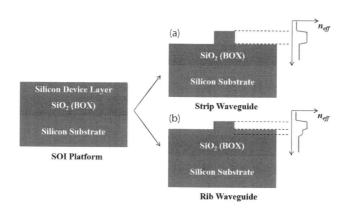

FIGURE 5.2
Cross-sectional structure of the SOI platform, and (a) strip and (b) rib waveguides.

the light signal is confined inside the waveguide core. Laterally there is no silicon slab surrounding the waveguide core. Since the effective refractive index n_{eff} of the waveguide core region is higher than that of the surrounding regions, which are typically air or SiO_2, the confinement of light in the silicon waveguide core is achieved.

On the other hand, in the rib waveguide, the waveguide core is laterally surrounded by a thin silicon slab. Therefore, the refractive index contrast between the waveguide core and the surrounding regions is lower than that in the strip waveguide, and the light confinement provided is weaker. This can be confirmed by the electric-field distribution (the lowest order mode) inside the strip or the rib waveguide. As shown in Figure 5.3, the electric-field is confirmed in a smaller region in the strip waveguide, especially in the lateral direction.

Since the intrinsic loss of silicon when the photon energy is smaller than the band gap is very low, the propagation loss of silicon waveguides is mainly due to the scattering caused by the sidewall roughness of waveguide core. The upper bound of the scattering loss α_m can be theoretically calculated as [6]:

$$\alpha_m = \frac{\sigma^2 \kappa}{k_0 d^4 n_1} \tag{5.1}$$

where σ is the root-mean square roughness, k_0 is the wavevector of light in vacuum, d is the half-width of the waveguide core, n_1 is the effective refractive index of a silicon slab with the same thickness as the waveguide core and κ is a factor depending on the waveguide geometry and the statistical distribution of roughness. Commonly, the sidewall roughness of rib waveguides is smaller than that of strip waveguides, and hence, the signal propagation loss of rib waveguides is lower. As slow as <0.3 dB/cm propagation loss has been achieved with the silicon rib waveguide [7], whilst the propagation loss of strip waveguide is normally in the order of 1–2 dB/cm [8].

In addition to straight waveguides, bending waveguides are also needed in the silicon photonic integration platform. For compactness of integrated devices and circuits, sharp waveguide bends with small radium is highly desirable. When the optical signal propagates through the waveguide bend, the major loss is the radiation loss. For rib waveguides, due to the relatively

Strip Waveguide   Rib Waveguide

FIGURE 5.3
Electric-field distribution of strip and rib waveguides.

weak light confinement, the radiation loss is usually higher than that of the strip waveguides. Therefore, strip waveguides are preferred in integrated structures with bends. With appropriately optimized geometry, <0.03 dB/ bend loss has been achieved [9].

Due to the rectangular shape, silicon waveguides are mostly polarization-dependent and have relatively large structural birefringence [10]. Depending the electric-field oscillation direction, silicon waveguides support transverse-electric (TE) and transverse-magnetic (TM) polarizations or modes. The electric-field distributions (cross-section) of the lowest order TE-mode and TM-mode are shown in Figure 5.4. It can be seen that the TE and TM modes oscillate in the vertical and lateral directions, respectively. Due to the lateral distribution, the TM mode generally has higher bending loss compared to the TE mode, when the bending radium is kept the same.

Depending on the waveguide geometry, different number of modes are supported in silicon waveguides. For the popular SOI platform with 220 nm thick top silicon layer, when the waveguide width is smaller than about 460 nm, only the fundamental TE mode (i.e. TE_0) and the fundamental TM mode (i.e. TM_0) are supported. The fundamental modes have the lowest waveguide propagation loss and bending loss, and such single-mode waveguide is normally used in silicon photonic integrated circuits to avoid mode crosstalk and coupling issues.

5.2.2 Silicon Integrated Polarization Splitter and Rotator

As discussed in the previous section, silicon waveguides have relatively large birefringence. Since almost all integrated devices are based on waveguide structures, silicon photonic integrated circuits also have large birefringence and their performance is polarization dependent. However, in practical applications, the input signal normally has arbitrary polarization, and hence, integrated polarization control devices are required.

The most widely used concept to deal with the polarization dependence limit of silicon photonic integrations is the polarization diversity [11].

TE₀ Mode **TM₀ Mode**

FIGURE 5.4
Electric-field distribution of TE_0 and TM_0 modes.

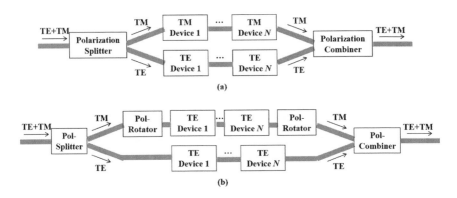

(a)

(b)

FIGURE 5.5
The principle of polarization diversity in silicon photonic integrations. (a) Polarization diversity with TE and TM optimized components; and (b) polarization diversity with polarization rotator and TE optimized components.

The basic principle of the polarization diversity scheme is shown in Figure 5.5a. In this scheme, the input signal is split into two channels with orthogonal polarizations (i.e. TE and TM, respectively). Signals with orthogonal polarizations propagate through the integrated functions separately, and they are combined at the output. The splitting and combination of orthogonal polarizations are realized using polarization splitters/combiners, which will be discussed in detail later. In addition, since most integrated photonic device are designed based on the TE mode, to avoid the re-design of devices, a polarization rotator can be inserted into the TM mode signal path directly after the polarization splitter, as shown in Figure 5.5b. The polarization rotator converts the TM polarization into the TE polarization, and hence, the same devices can be used in the TM signal path. Right before the polarization combiner at the output side, another polarization rotator is used in the TM signal path to convert the signal back to the TM polarization.

The polarization splitter/combiner can be realized using a number of principles, and one of the most widely used schemes are based on the directional coupler structure [12]. As shown in Figure 5.6, with appropriately designed waveguide geometry, the coupling length L_c (i.e. the length of coupler where

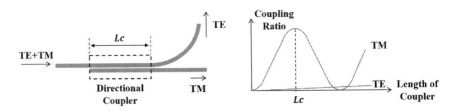

FIGURE 5.6
Directional coupler-based silicon integrated polarization splitter/combiner.

the signal is fully coupled to the other waveguide) of the TM mode is significantly shorter than that of the TE mode. Therefore, by controlling the length of directional coupler, the TE and TM polarizations can be split. In addition to the directional coupler-based polarization splitter/combiner, a number of other principles have also been proposed and demonstrated, such as the 2D grating coupler- and the multi-mode interferometer (MMI)-based schemes [13,14].

In addition to polarization splitters/combiners, the silicon integrated polarization rotators are also highly demanded. The rotation of optical axis is required to change the signal polarization, which is primarily realized by introducing asymmetrical structures, such as using off-axis double cores, bi-level tapers and cascaded bends [15–17]. One example of silicon integrated polarization rotator is shown in Figure 5.7, which is based on an adiabatic taper and an asymmetrical direction coupler [18]. The bottom cladding of the structure is SiO_2 and the top cladding is air, and hence, the vertical symmetry of the structure is broken. When light with both TE_0 and TM_0 polarizations propagates through the adiabatic taper, the fundamental TE mode is unchanged, whilst the fundamental TM mode launched at the narrow side is converted to the first higher order TE mode (i.e. TE_1 mode) at the wide side due to the mode coupling. Then the two modes propagate through the asymmetric coupler, which consists of a narrow waveguide and a wide waveguide. The effective refractive index of the TE_1 mode in the wide waveguide is the same as that of the TE_0 mode in the narrow waveguide, and hence, by selecting the length of asymmetric coupler appropriately, the TE_1 mode input from the wide waveguide is coupled to the TE_0 mode that outputs from the narrow waveguide. At the same time, the TE_0 mode input from the wide waveguide is not changed in the asymmetric coupler and hence, it propagates out from the wide waveguide directly. Through this way, the two polarizations of the input signal are separated, and the TM polarization is converted to the TE polarization simultaneously. Therefore, the polarization diversity principle shown in Figure 5.5b can be realized.

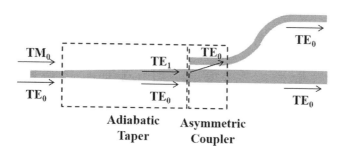

FIGURE 5.7
Adiabatic taper and asymmetric coupler-based polarization rotator.

5.2.3 Silicon Integrated Optical Filters

In optical communication systems, such as the near-infrared indoor OWC systems discussed in this book, optical filters, mostly bandpass filters, are widely used to select the optical signal with desired wavelength and to reject other out-of-band signals. Therefore, to realize integrated transceivers for indoor OWC systems, silicon integrated optical filters are compulsory.

Silicon integrated optical filters can be realized based on a number of options, and they rely on the interference of optical signals in general. One widely used principle for silicon integrated optical filters is the Mach–Zehnder interferometer (MZI), as shown in Figure 5.8. The input signal is split into two parts and they propagate through two arms of the MZI. Generally, the waveguides in two arms are identical, and the total length of these two arms is L_1 and L_2, respectively. Assume the effective refractive index of the silicon waveguide is n_{eff}, then the phase difference $\Delta\varphi$ of the signals passing through the two arms can be expressed as:

$$\Delta\varphi = \frac{2\pi n_{eff}}{\lambda}\left(L_2 - L_1\right). \tag{5.2}$$

Since n_{eff} is typically a function of signal wavelength, signals from two arms interference constructively at some wavelengths and destructively at other wavelengths at the output. Therefore, the optical filtering function is realized, and the filter spectral is similar to a sine wave. In addition to having identical waveguides in both arms, different waveguide widths can be used to control the phase difference more effectively. Multiple MZI stages can be cascaded as well to realize other filter spectral characteristics [19].

One major limitation of MZI-based silicon integrated optical filters is the relatively long arms required for a narrow passband, which results in large device footprint. To realize compact size and dense integrations, micro-ring resonators have been proposed and widely studied. The compact size of micro-ring resonators is enabled by the large refractive index contrast between the silicon waveguide core and surrounding claddings, which

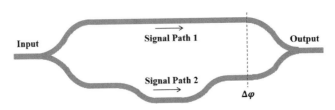

FIGURE 5.8
MZI-based optical filter in silicon photonic integrations.

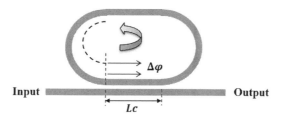

FIGURE 5.9
Micro-ring resonator-based optical filter in silicon photonic integrations.

provides tight light confinement and allows sharp waveguide bends. Both circle rings and racetrack-shaped rings can be used. A racetrack-shaped ring resonator-based integrated filter is shown in Figure 5.9. The ring is essentially a loop optical waveguide, and the phase delay of optical signal with a wavelength λ after one round trip can be expressed as:

$$\Delta\varphi(\lambda) = \frac{2\pi n_{eff}(\lambda)}{\lambda}\left(2\pi r + 2L_C\right) \qquad (5.3)$$

where r is the bend radium and L_C is the coupler length. When the phase delay after one round trip satisfies $\Delta\varphi = m \cdot 2\pi$, where m is an integer, the signal experiences constructive interference inside the ring resonator. Therefore, the wavelength satisfying the phase matching condition has low transmission coefficient at the output, whilst other wavelengths can pass through to the output. Therefore, the optical filter function is realized.

Compared to MZI-based silicon integrated optical filters, the micro-ring-based solution has much smaller size (several μm^2) and narrower passband, whilst the major limitation is the smaller free-spectral range (FSR) [20]. To further reduce the passband of silicon integrated optical filters, higher-order ring resonators can be used, where integrated optical bandpass filters with GHz scale narrow bandwidth have already been demonstrated [21]. In addition to the MZI- and micro-ring resonator-based schemes, silicon integrated optical filters can also be realized using other principles, such as waveguide Bragg gratings and arrayed waveguide gratings [22,23].

5.2.4 Silicon Integrated Optical Power Splitters

In silicon photonic integrated circuits, the splitting of input signal to multiple parts is widely utilized, such as for signal interference in the MZI-based optical filters shown in Figure 5.8. Therefore, optical power splitters are highly demanded.

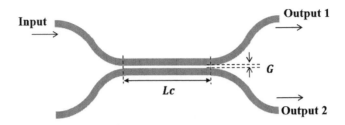

FIGURE 5.10
Silicon integrated optical power splitter based on the directional coupler.

The directional coupler principle is the most common power splitting method in optical fiber communication systems, and it is also popular in the silicon photonic integration platform. The directional coupler consists of two parallel silicon waveguides, as shown in Figure 5.10. Both strip and rib waveguides can be used in the directional coupler, and here we use the strip waveguide-based device as an example. Due to refractive index matching, the optical signal is coupled between two waveguides. The behavior of directional coupler can be analyzed using the coupled mode theory [24]. Based on the theory, the ratio of power coupled from one waveguide to the other can be calculated as:

$$\eta = \frac{P_{coupled}}{P_0} = sin^2 \left(C \cdot L_C \right) \tag{5.4}$$

where P_0 is the power of input signal, $P_{coupled}$ is the power coupled to the other waveguide, C is the coupling coefficient and L_C is the coupler length. The coupling coefficient is mainly controlled by the width of waveguides and the spacing G between two silicon waveguides, and it can be analyzed using the "supermode" method, which is based on numerical calculations. The coupler length can then be selected to have 50% of original power coupled to the other waveguide to realize the power splitter function.

In addition to the directional coupler, the Y-branch and the multi-mode interferometer (MMI) are also widely used for the silicon integrated optical power splitter [25,26]. Ordinarily, we characterize the performance of silicon integrated optical splitter with three major parameters. The first one is the excess loss, which is the difference between the total output power and the input power. The second one is the operation bandwidth, and broadband bandwidth is normally desirable to enable the device to be used over a large wavelength region. The third key parameter is the polarization dependence. Polarization independent power splitters can be used with arbitrary input signal, whilst polarization splitter and rotator discussed in Section 5.2.2 are needed for polarization-dependent power splitters.

In addition to the basic silicon waveguide, the polarization splitter and rotator, the optical filter and the optical power splitter discussed in this section, there are more types of passive devices that are required and demonstrated on the silicon photonic integration platform, such as optical reflectors and contra-directional couplers [27,28]. Interested readers can refer to reference [29] for more detailed introductions and discussions.

5.3 Silicon Photonic Integrated Active Devices

The devices discussed in Section 5.2 do not require external power for operation, and hence, they are referred as passive optical devices. In addition to basic passive devices, active devices are also important in silicon photonic integrations, such as optical modulators, detectors and switches. Active devices actually provide more control capability to silicon photonic integrated circuits, and they enable more advanced functionalities. For example, with optical switches, dynamic reconfiguration of the optical transceiver or system becomes possible. Active devices also provide the tuning capability, such as realizing tunable optical bandpass filters to realize the reconfigurable optical add-drop multiplexer (ROADM). In this section, we briefly introduce the most widely used silicon photonic integrated active devices (i.e. the modulator and the PD).

5.3.1 Silicon Integrated Optical Modulators

The optical modulator is a device to modulate the light signal according to the applied electrical signal. In traditional optical communication systems using discrete components, optical modulators are usually realized using $LiNbO_3$ through the electro-optic effect. However, silicon does not have the electro-optic effect, and hence, other principles need to be explored to realize silicon photonic integrated optical modulators.

Typically, silicon integrated optical modulators are realized using either the plasma dispersion effect or the thermal-optic effect. The plasma dispersion effect refers to the property of silicon that the refractive index changes according to the carrier density. Based on experimental data, the change of refractive index of silicon in the mostly widely used 1550 nm telecommunication band can be calculated as [30]:

$$\Delta n = -5.4 \times 10^{-22} \Delta N^{1.011} - 1.53 \times 10^{-18} \Delta P^{0.838} \tag{5.5}$$

where ΔN and ΔP are carrier densities of electrons and holes in the unit of cm^{-3}, respectively. PN junctions are normally used for the controllable change of carrier densities.

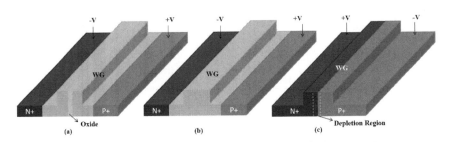

FIGURE 5.11
Silicon integrated optical modulators. The modulators are based on: (a) carrier accumulation;
(b) carrier injection; and (c) carrier depletion.

Based on the plasma dispersion effect and PN junction, optical phase
modulators can be realized, where the phase of optical signal changes
according to the applied electrical signal. The change of signal phase is
achieved by changing the carrier density, and the electrical manipulation
of charge carrier density can be realized through carrier accumulation,
carrier injection and carrier depletion. The basic principles of these three
schemes are shown in Figure 5.11. In the carrier accumulation-based
scheme, a thin layer of SiO_2 is used between the two halves of silicon
waveguide to provides insulation. Therefore, when positive and negative voltages are applied to the P and N regions, respectively, the waveguide works like a capacitor, where a large number of carriers accumulate
around the SiO_2 insulation layer. The number of accumulated carriers
can be changed by varying the applied voltages, and hence, the refractive index of the silicon waveguide can be changed accordingly through
the carrier plasma effect. Therefore, when the light signal propagates
through the waveguide, its phase can be controlled and the device is a
phase modulator.

In addition to carrier accumulation, the carrier injection method can also
be used to realize silicon phase modulators. The basic principle is shown
in Figure 5.11b. Highly doped *p*- and *n*-regions are separated by the intrinsic waveguide region. By applying positive and negative voltages to the *P*
and *N* regions, respectively, carriers are injected into the waveguide region.
Hence, the refractive index of silicon waveguide is changed and the phase of
light after passing through the waveguide can be controlled by the applied
voltages.

The third major method based on the plasma dispersion effect used for
optical phased modulation in silicon is shown in Figure 5.11b, which is based
on carrier depletion. Here, in addition to the highly doped *p*- and *n*-regions,
the waveguide is also lightly *p*- and *n*-doped. Due to the PN junction inside
the waveguide, there is a depletion region formed. By applying reverse
bias (i.e. applying positive and negative voltages to the *N* and *P* regions,
respectively), the width of depletion region changes accordingly. Therefore,

the carrier density inside the waveguide varies and the refractive index of the waveguide is modulated. When the light signal propagates through the waveguide, phase modulation is realized.

The carrier accumulation, injection and depletion methods have all been explored in silicon integrated optical modulators [5]. Moreover, due to the use of capacitor-like principle, the modulation speed supported is the lowest in the carrier accumulation-based modulators, whilst the carrier depletion-based solution provides the best modulation speed performance. On the other hand, since the carrier depletion scheme has the largest number of carriers available for the plasma dispersion effect, it provides the largest change of refractive index, and hence, the best modulation efficiency. The carrier depletion-based scheme can only change the refractive index slightly, and the modulation efficiency is usually limited. Due to this, carrier depletion-based silicon optical modulators typically require long waveguide and large footprint to achieve reasonable modulation depth.

The optical modulators discussed above are only capable of changing the phase of optical light through the plasma dispersion effect. In optical communication systems, especially the indoor OWC systems, instead of phase modulation, the IM is more widely used. To realize silicon integrated optical intensity modulator, the MZI structure can be used. In the MZM, as shown in Figure 5.12, the input signal is first split into two parts. The first part directly propagates through a silicon waveguide to the output, whilst the second part goes through another waveguide with a phase tuning region. The phase tuning can be based on carrier accumulation, carrier injection, or carrier depletion, as introduced above. Then the two parts of light signal interfere with each other at the output. By controlling the phase change induced by the phase tuning region, the interference of the two parts of signal can be either constructive or destructive, which realizes the IM. With the carrier depletion-based phase tuning method, over 40 Gb/s silicon integrated optical MZM has been demonstrated [31].

In addition to using the plasma dispersion effect to introduce voltage dependent refractive index change, the thermal-optic effect has also been investigated to realize the refractive index change, and hence, silicon

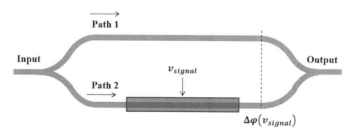

FIGURE 5.12
Silicon integrated optical intensity modulators based on the MZI structure.

integrated optical modulators. In the thermal-optic effect-based phase shifter, a metal resistor is generally deposited above the silicon waveguide. By applying signal dependent voltage to the metal resistor, heat is generated, and the heat descends towards the waveguide. Therefore, the refractive index of the silicon waveguide is changed, and the phase tuning of light passing through the waveguide is achieved. By further using the MZI structure shown in Figure 5.12, thermal-optic effect-based optical intensity modulator is realized. Since the silicon material has a relatively high thermal-optic coefficient, this type of modulators primarily have high modulation efficiency. However, compared to the plasma dispersion effect, the thermal-optic effect has a longer time constant. Therefore, the speed of thermal-optic effect-based modulators is usually limited [32], and the plasma dispersion effect-based modulators are more widely used in the silicon photonic integration platform.

The MZM structure discussed above is widely used in silicon integrated optical modulators, and the size of MZM is determined by the phase changing efficiency of the phase tuning region. Since the plasma dispersion effect is a relatively weak effect, the phase tuning region normally needs to be long (in the range of hundreds of micrometers to millimeters) and the resulting MZM occupies a large space in the integrated photonic circuit. To solve this issue, the micro-ring resonator-based structure has been widely explored to realize compact silicon integrated optical modulators. This is based on the wavelength selectivity of micro-ring resonators, where the resonant wavelength is determined by the signal round-trip phase delay. Therefore, by changing the refractive index of the micro-ring resonator through the plasma dispersion effect, the resonance wavelength of the cavity can be changed. As shown in Figure 5.13, when the operation signal wavelength is close to the resonance wavelength, the transmission is very sensitive to the phase change of the cavity. As a result, even with a small phase change in

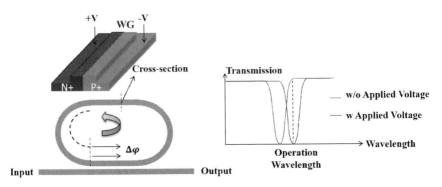

FIGURE 5.13
Silicon integrated optical intensity modulators based on the micro-ring resonator.

the micro-ring cavity, the transmission of output signal varies significantly. Therefore, a highly efficient optical intensity modulator can be realized using the micro-ring resonator. Based on the micro-ring resonator structure and carrier depletion-based plasma dispersion effect, over 50 Gb/s silicon integrated optical intensity modulator with low operation voltage and ultra-compact size has been demonstration [33].

5.3.2 Silicon Integrated Photodetectors

In optical communication systems, the data to be transmitted is modulated onto the optical carrier at the transmitter side, and the optical signal is converted back to the electrical domain at the receiver. PDs are generally used for the O-E conversion, and for high-speed operations, the PIN photodiode is most widely used. In this section, we introduce silicon integrated PIN PDs for optical communications.

The operation of PDs is based on the absorption of incident photons, and as discussed in Section 2.2.2, the condition of photon absorption is:

$$\frac{hc}{\lambda} \geq E_g \qquad (5.6)$$

where h is the Boltzmann constant, c is the speed of light in vacuum, λ is the light wavelength and E_g is the bandgap of PD material. Since silicon has a relatively large bandgap of 1.12 eV, it can only detect light with the wavelength shorter than about 1100 nm. Therefore, silicon cannot be used for photodetection in typical communication systems directly, which operate in the 1550 nm band.

From the simplicity perspective, it is highly desirable to use silicon as the absorption material of PD directly. Whilst bulk silicon cannot be used in the 1550 nm band directly, several techniques have been developed, such as the damage-based detectors [34]. However, these solutions still suffer from low responsivity, high dark current and limited operation bandwidth. Therefore, they are not practical in telecommunication applications.

To solve this issue, CMOS-compatible materials that can be integrated together with silicon have been explored for 1550 nm silicon integrated PDs. The most widely used solution is the germanium-on-silicon (Ge-on-Si) PD, where the germanium is deposited on top of silicon for the absorption of input light, as shown by the cross-section in Figure 5.14. Here the light is absorbed in the Ge layer, which generates electron and hole pairs. Reverse bias is applied to the structure, and hence, electrons and holes drift through the *n*-doped and *p*-doped regions, respectively. Therefore, photocurrent is generated externally, and the input signal is detected successfully.

In Ge-on-Si PDs, the light signal can incident to the structure in two ways. The first is vertical incidence. Therefore, due to the finite absorption coefficient, the Ge layer needs to be relatively thick to absorb the input signal effectively.

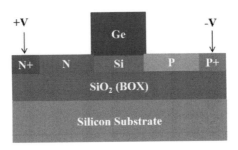

FIGURE 5.14
Silicon integrated PD.

However, it takes a long time for photon-generated electrons and holes to move out of the thick Ge layer. Hence, the operation speed is highly limited. To solve this trade-off, the second light incidence method is proposed, where the light signal propagates in a silicon waveguide that is underneath the Ge absorption layer. In this way, the optical signal is evanescently coupled to the Ge layer while propagating along the waveguide, and a thin layer of Ge is sufficient to achieve highly efficient photon absorption. Therefore, both high-speed operation and high responsivity can be achieved simultaneously. To further increase the operation bandwidth, the distribution of electric-field inside the integrated PD can be further optimized. Based on this principle, over 40 GHz Ge-on-Si integrated PIN photoreactor with over 0.8 A/W responsivity has been demonstrated [35].

In this section, we have briefly introduced two types of most widely used silicon integrated active devices. In addition to the modulator and the PD, other active devices are also available, and most of them are based on the phase tuning/modulation principle discussed in Section 5.3.1. One example is the optical tunable bandpass filter, which can be realized based on either the MZI structure or the micro-ring resonator. By adding a phase tuning region to one arm of MZI, the wavelength satisfying the constructive (and destructive) interference condition can be changed according to the applied voltage. Therefore, the spectrum of bandpass filter is tunable. Similarly, by using the micro-ring resonator with a phase tuning region, the resonance wavelength can be changed by applying an external voltage, and the central wavelength of the filter becomes tunable. Such principle can also be applied to realize optical switch in the silicon photonic integration platform.

Although a large number of active functions and devices can be realized using the silicon photonic integration technology, integrated lasers are still challenging fundamentally. Silicon is an indirect-band semiconductor, and hence, it cannot provide optical gain and is inefficient for light generation. Due to this constrain, the light source for silicon photonic integrated circuits is mainly based on external lasers or hybrid silicon lasers. The hybrid silicon laser is based on the heterogeneous integration (e.g. bonding) of active III–V materials (such as InP) onto the silicon platform. Compared to external lasers, the hybrid laser solution is more attractive, since it enables more efficient laser light coupling and better integration. The use of Ge to realize silicon integrated laser has also been proposed and studied. Whilst the silicon integrated laser has seen significant progress in recent years, it still requires more study to further optimize the performance for practical applications, such as reduce the threshold current, increase the efficiency, and simplify and optimize the fabrication.

5.4 Silicon Photonic Integrations for Indoor OWC Systems

In the previous sections, we have discussed the high demand of photonic integration technologies in optical communication systems and provided the introduction of the popular silicon photonic integration technology. In this section, we will focus on the use of photonic integration in near-infrared high-speed indoor OWC systems, to realize integrated compact transceivers and systems with low-cost and high performance.

Compared to traditional optical communication systems where the light signal propagates inside optical fibers, the light signal transmission medium is the free space in indoor OWC systems. The use of free space for signal transmission has two major limitations, that is, the loss of beam confinement and the vulnerability to environmental interference (e.g. background light). On the other hand, unlike the optical fiber link, where light can only propagate along the pre-defined path, the free-space transmission link provides the advantage of flexibility, since the light signal propagation path can be reconfigured adaptively.

This unique advantage is explored in the beam steering-based indoor OWC system discussed in Section 2.3. In this type of system, the free-space propagation direction of data-carrying optical signal is steered dynamically in the system according to the location of user, and this beam steering function allows achieving high-speed communication and low transmission power simultaneously. Therefore, the beam steering device is unique and plays a critical role in high-speed indoor OWC systems. In this section, we will focus on the integration of this unique device using the silicon photonic technology. We will review currently popular beam steering principles and devices in Section 5.4.1; introduce the optical phased array principle for beam steering without mechanical movement in Section 5.4.2; and discuss the silicon integrated beam steering device based on the phased array principle in Section 5.4.3.

5.4.1 Beam Steering Devices in Indoor OWC Systems

As discussed above, the beam steering device allows the dynamic tuning of light propagation direction in the free space according to practical requirements, and hence, it is highly demanded in high-speed indoor

OWC systems. A number of technologies have been studied to realize beam steering devices, and they mainly include the micro-electromechanical system (MEMS), the liquid crystal and the optical phased array principles. We will review the MEMS- and the liquid crystal-based schemes in this section, and introduce the optical phased array principle in the next section.

MEMS is a miniaturization technology that combines mechanical and electronic structures, components and functions. It commonly consists of the micro-sensor, the micro-actuator, the micro-electronics and the micro-structure sections. The micro-structure defines the basic mechanical structure, the micro-sensor converts a measurable mechanical signal to an electrical signal, the micro-electronics process the electrical signal and control the mechanical structure, and the micro-actuator provides the mechanical force to move the microstructure.

One key device realized using the MEMS technology is the steering mirror, where a metal-coated reflector can be steered by applying control voltages using the micro-actuator. To drive the actuator and to steer the mirror, the use of piezoelectric effect and electro-static force has been proposed and demonstrated [36,37]. With the popularity of CMOS fabrications, silicon-based MEMS devices have attracted substantial interest, and 2D steering mirrors have been realized [37].

The use of MEMS-based steering mirrors in indoor OWC systems have been proposed and demonstrated as well [38]. The major advantages of using MEMS steering mirrors for the beam steering function mainly include: (1) the insertion loss of the beam steering device is low: with appropriate coating of the mirror, over 99% reflectivity can be achieved; (2) the operation principle is simple: MEMS mirrors are based on optical signal reflection, and hence, the free-space light propagation direction can be easily controlled; (3) 2D continuous beam steering is possible, which enables full mobility of the user in the indoor OWC system; and (4) the steering range is relatively large: over 20° angular steering range has been demonstrated [37], allowing a large signal coverage area.

However, MEMS steering mirrors for the beam steering in indoor OWC systems also have several limitations. First, MEMS mirrors usually require a high operation voltage (e.g. over 10 V), especially when a large steering range is required. This may lead to the safety concern and the high power consumption issue. Second, the steering speed of MEMS mirrors is limited, and there is a trade-off between the steering range and the speed. Normally, at least several milliseconds are required for the point-to-point steering, and the steering time is even longer when a large steering range is required. Third, the MEMS mirror has a non-planar structure, and hence, it is difficult to be integrated together with other components. Therefore, it is challenging to use MEMS steering mirrors in photonic integrated circuit with multiple functions and devices. In addition, the steering stability and the steering accuracy of MEMS mirrors are also limited.

In addition to MEMS mirrors, the liquid crystal principle can also be used to achieve the beam steering function in indoor OWC systems, especially the liquid-crystal-on-silicon (LCoS) device. In LCoS, micro-sized liquid crystals are formed on the silicon platform. When the light signal incidents onto the liquid crystal (i.e. pixel), the light is diffracted. The diffracted lights from multiple pixels then experience interference, and the interference pattern is created. By controlling the diffraction of each pixel, which is generally realized by changing the shape of liquid crystal, the interference pattern can be controlled. Through this way, constructive interference can be generated in a certain direction, and this direction can be controlled adaptively. Therefore, the beam steering function can be achieved in indoor OWC systems.

LCoS in OWC systems is also known as the spatial light modulator (SLM). The application in indoor OWC systems has several challenges, which mainly include: (1) the operation of LCoS is based on signal diffraction and interference from a large number of pixels, and hence, the control of beam steering is relatively complicated; (2) the insertion loss of LCoS is relatively high, especially when the beam steering angle required is large, since lossy high-order diffraction signals need to be used. This results in additional loss in the power-limited indoor OWC system and (3) the cost of LCoS (i.e. SLM) is normally high, and it is also difficult to be integrated with other optical devices, similar with MEMS steering mirrors.

Due to the limitations of MEMS steering mirrors and LCoS, the optical phased array principle with waveguide structures has been explored to achieve the beam steering function in indoor OWC systems. Optical phased arrays with waveguide structures (referred as optical phased arrays in this book) become even more attractive when considering the system and transceiver integration demands in OWC systems and the rapidly developing silicon photonic integrated circuits, since they have a planar structure and are compatible with other devices. Therefore, we will focus on the optical phased array principle and silicon integrated optical phased arrays for beam steering in indoor OWC systems in the next section.

5.4.2 Optical Phased Arrays for Beam Steering

The phased array principle has been widely studied and used in the microwave field, and phased arrays have replaced the conventional horn antennas in many cases. This is mainly because of the advantages of random-access and rapid microwave beam forming without moving parts. Due to these strengths, phased arrays are popular choices in RADAR applications, which require the steering of microwave sensing signals.

The phased array principle has also been extended to the optical frequency range, and the resulting technique is normally called the optical phased array. Similar with microwave phased arrays, optical phased arrays have also been widely used for optical beam forming and steering. For example, optical phased arrays have been widely used in LiDAR, which is the optical version of RADAR, for target detection, wind profiling and gas cloud identification. Optical phased arrays have also been used in space laser communications, to provide high-speed wireless data connections between the satellite and the ground or between satellites. With the rapid development of photonic integration technologies, especially the silicon photonic integration platform, the miniaturization and integration of optical phased arrays become possible. Therefore, it is promising to use low-cost and small-sized silicon integrated optical phased arrays for the beam steering function in high-speed indoor OWC systems. We will review the basic optical phased arrays principle here, and introduce the silicon integrated optical phased array in the next section.

For simplicity, we use a one-dimensional (1D) optical phased array as an example, as shown in Figure 5.15. Each phased array element consists of an optical antenna (e.g. the output facet of silicon waveguide) that generates the diffraction pattern in the free space, and a phase tuning region that controls the phase difference in the optical phased array. Assume the total number of elements in the optical phased array is N, the distance between the antennas of adjacent phased array elements is d and the relatively phase delay between adjacent phased array elements is φ. As shown in the figure, the lights diffracted by the optical antennas interfere with each other and creates an interference pattern at a distance L from the optical phased array. When the diffracted signals interfere constructively, a peak will appear in the interference pattern. Assume the size of the optical phased array is much smaller than the distance L (i.e. $N \cdot d \ll L$), then the angle θ between the optical phased array and the peak of the interference pattern (referred as the main lobe) can be calculated as [39]:

$$\sin(\theta) = \frac{\lambda_0 \varphi}{2\pi d} \tag{5.7}$$

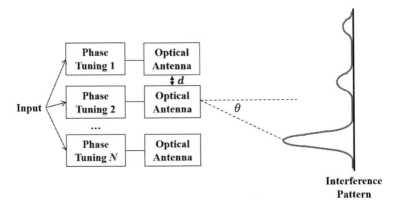

FIGURE 5.15
The basic principle of optical phased array.

where λ_0 is the free-space wavelength of the signal. As shown by the equation, by changing the relative phase delay φ between adjacent phased array elements, the angle of light interference peak can be varied accordingly. Consequently, the optical beam generated by the optical phased array can be steered in the free space.

In addition to the beam steering angle, another important parameter in indoor OWC systems is the divergence of the beam generated by optical phased array, since it will decide the beam footprint (i.e. service coverage area) after propagating through the free space. The beam divergence can be calculated as [39]:

$$\Delta\theta \approx \frac{0.886\lambda_0}{Ndcos(\theta)} \tag{5.8}$$

where $\Delta\theta$ is the full-width half-maximum (FWHM) divergence angle of the generated optical beam. It is clear from Eq. (5.8) that a small beam divergence can be achieved by using a larger number of elements in the optical phased array or increasing the distance between adjacent phased array elements. However, a large distance between adjacent elements reduces the maximum beam steering angle, and hence, the trade-off needs to be considered when designing the optical phased array.

In addition, as can be seen from Figure 5.15, in addition to the main lobe with the highest interference peak, side lobes may also be generated by the optical phased array. The existence of side lobes results in the power radiation in unwanted directions. Therefore, the power efficiency of the desired main lobe is reduced, and crosstalk with be induced in the system. The side lobes also reduce the effective beam steering range, since the beam cannot be steered beyond the space of two lobes. Here we define the angular spacing between the main lobe and the first side lobe as ϕ, and it can be calculated from the phase condition of constructive interference, which can be expressed as:

$$sin(\phi) = \frac{\lambda_0}{d} \tag{5.9}$$

It is clear from Eq. (5.9) that a small spacing between adjacent phased array antennas is needed to increase the angular space between the main lobe and the side lobe, which is required to increase the supported beam steering range. In addition, when spacing between adjacent phased array antennas approaches the half-wavelength, a full 180° beam steering range can be achieved. In practice, the spacing between antennas is limited by the footprint requirements of other building blocks of the optical phased array (e.g. phase tuning regions) and the crosstalk between multiple signal channels in the array.

5.4.3 Silicon Integrated Optical Phased Arrays

Due to the unique advantages of optical phased arrays, especially the possibly of realizing waveguides-based structures, the use of optical phased array for the beam steering function is promising. With the integration efforts and trends of indoor OWC systems, integrated optical phased arrays are also highly demanded. In this section, we discuss the recent developments of silicon integrated optical phased arrays.

As shown by Figure 5.15, three major functions are required in silicon integrated optical phased arrays (i.e. power splitters, phase tuning regions and optical antennas). The power splitter divides the input signal into multiple coherent parts for the propagation through phased array elements. The phase tuning region controls the phase delay introduced to each element, and hence, determines the relative phase delay φ amongst adjacent channels inside the array. The optical antenna serves as the interface from the integrated phased array to the free space to create the interference pattern and to realize the beam steering function.

Silicon integrated optical power splitters have been discussed in Section 5.2.4, and they are normally realized based on directional coupler or MZI, MMI and Y-branch structures. However, the previously proposed and demonstrated silicon integrated power splitters have several limitations. Whilst Y-branches have the advantage of simple structure, the insertion loss is normally high due to mode scattering, or a long device length is required to achieve a reasonably low loss. Directional couplers can solve the high loss and long device issues. However, they usually have high wavelength and polarization dependence, which results in a relatively narrow bandwidth and only supporting one specific signal polarization. In addition, the power splitting ratio is sensitive to fabrication variations. The MMI structure is capable of supporting a larger bandwidth with higher fabrication tolerance and smaller size, whilst the bandwidth is still limited to <100 nm and it is also polarization-dependent. To solve these issues, here we introduce a highly compact, broadband and low-loss silicon integrated optical splitter supporting arbitrary signal polarization [40]. This power splitter is based on silicon adiabatic tapered waveguides, and it is ideal for the beam steering function in indoor OWC systems using silicon photonic integrations.

The basic structure of the silicon integrated optical splitter is shown in Figure 5.16. It mainly consists of three regions, which are the input waveguide, the tapered waveguides and the output waveguides. Three tapered waveguides are used with one input taper and two output tapers, and the output waveguides are bending waveguides to separate the two output signals. In the power splitter, a single-mode input strip waveguide with a dimension of $W \times 220$ nm is gradually tapered down to a tip width T along a length L, where W is the original waveguide width and 220 nm is the height of waveguide (i.e. the top device silicon layer thickness is 220 nm). The input tapered waveguide is symmetrically inserted into two output tapered waveguides,

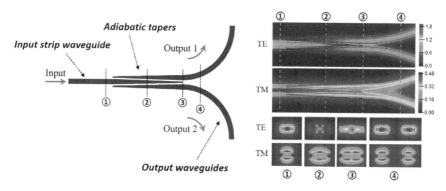

FIGURE 5.16
Structure of silicon integrated optical power splitter based on adiabatically tapered waveguides.

and the gap between two adjacent tapers is selected as G. The output tapered waveguides have the same length (L), waveguide width (W) and tip width (T) as the input tapered waveguide. The two output tapered waveguides are then connected with the two bending output waveguides.

In the proposed splitter, inside the tapered region, the input light is adiabatically coupled from the middle waveguide to two symmetric outside waveguides through three tapers. As long as the adiabatic mode evolution is satisfied for both TE and TM polarizations, the 3 dB power splitting ratio can be achieved for both polarizations at the end of the tapered region. Then the signal after propagating through the tapers is split into the fundamental modes through the two bending waveguides to the outputs.

The mode evolution process of both TE and TM polarizations is shown in Figure 5.16 (in different color scales), where the electric-field distribution simulated using the finite difference time domain (FDTD) method is shown. When the TE mode is launched, due to the electric-field discontinuities inside the tapered region, the light field is enhanced and concentrated within the gaps between input and output tapers. Along a certain propagation distance, the optical light can be seen as a lossless eigenmode propagating in a double-slot waveguide [41]. On the other hand, when the TM mode is launched into the device, as the electric-field is mainly distributed in the vertical direction at the top and bottom interfaces of the waveguide, the light is not squeezed into the gaps throughout the tapered region. Instead, the light with TM polarization propagating through the structure can be seen as a mode evolution process, where the input mode is evolved and coupled to two output waveguides adiabatically. Therefore, the proposed silicon integrated optical power splitter supports both signal polarizations.

In the adiabatic tapers-based silicon integrated optical power splitter, the geometric parameters, including the tapered waveguide length (L), original (or ending) width (W) and tip width (T), and the gap between adjacent tapered waveguides (*G*) need to be optimized. The detailed optimization process can be found in reference [40], and results have shown that a length (*L*) of as short as 5 μm is sufficient to achieve low-loss, broad operation bandwidth and polarization independence simultaneously.

The device, which is an important part of silicon integrated optical phased array for the beam steering in indoor OWC systems, has been fabricated using CMOS-compatible processes. An 8 inch SOI wafer with a 220 nm thick silicon layer and a 2 μm thick BOX layer is used, and the device is fabricated through e-beam lithography (EBL), E-beam evaporation, reactive-ion etching (RIE) and plasma enhanced chemical vapor deposition (PECVD) processes. The scanning electron microscope (SEM) image of the fabricated silicon integrated optical power splitter is shown in Figure 5.17. Measurement results show that an insertion loss of <0.19 and <0.14 dB can be achieved for the TE and TM polarizations, respectively, over the entire measured bandwidth from 1530 to 1600 nm.

The second major part of silicon integrated optical phased array is the phase tuning region. As discussed in Section 5.3.1, the phase shifter on silicon can be realized using either the carrier plasma dispersion effect or the thermal-optic effect. Since the beam steering function in the optical phased array is achieved via changing the relative phase delay, the steering speed is limited by the phase tuning region. Therefore, the plasma dispersion effect-based phase shifters are preferred for fast beam steering. However, due to the weak change of refractive index, long phase shifters are needed, resulting in less compact devices. Similar with optical modulators, the micro-ring resonator structure can be used to enhance the refractive index change caused by the plasma dispersion effect, and the phase shifter footprint can be reduced [42]. The cost of using the micro-ring structure is the reduced operation bandwidth. As a result, when compact phased arrays are required, the thermal-optic effect is usually used.

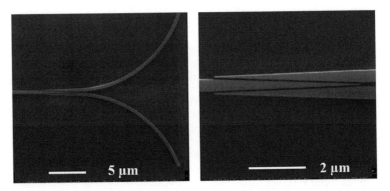

FIGURE 5.17
The SEM image of fabricated silicon integrated optical power splitter.

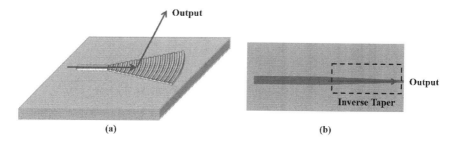

FIGURE 5.18
Antenna structure of silicon integrated optical phased array. (a) Grating coupler (perspective view) and (b) edge coupler (top view).

The third major part of optical phased arrays is the antenna, which radiates the optical signal from the photonic integrated circuit into the free space. The coupling of light out of silicon photonic integrated circuits is mainly realized via two devices (i.e. the grating coupler and the edge coupler). The basic structures are shown in Figure 5.18. The grating coupler-based antenna has a periodic structure, and the light is diffracted from the silicon waveguide in the horizontal direction to the free space in the vertical direction. On the other hand, the inverse taper is used in the edge coupler-based antenna, and the light is radiated into the free space along the original propagation direction (i.e. the horizontal direction). The inverse taper reduces the silicon waveguide width to squeeze the tightly confined optical filed out from the waveguide core to the surrounding cladding. This can reduce the mode mismatching between the waveguide mode and the free-space mode, and high radiation efficiency of the antenna can be achieved.

Silicon integrated optical phased arrays have been developed using both types of antennas. Due to the vertical direction out-of-chip radiation property of grating couplers, the grating coupler-based antennas can be arranged in two dimensions on the surface of silicon integrated photonic circuit. Therefore, grating couplers-based silicon integrated optical phased array have attracted intensive interests, and 2D beam steering has been realized [43]. In addition to organize the phased array antennas in two dimensions, the use of 1D phased array for 2D beam steering has also been developed [44]. The beam steering in one direction is achieved using the optical phased array principle discussed above via controlling the relative phase delay, whilst the beam steering in the other direction is realized using wavelength tuning. This is due to the periodic structure of grating couplers, where the out-of-chip coupling angle depends on the operation wavelength.

However, the grating couplers-based silicon integrated optical phased arrays also have several fundamental limitations for the beam steering application. First, the free-space radiation power efficiency of the grating coupler-based antenna is limited. This is mainly because that the light signal is coupled out from the integrated circuit into both upward and downward directions symmetrically, whilst only the upward coupled light can be used for practical OWC systems. Therefore, the typical power efficiency (only considering the out-of-circuit coupling into the free space) is limited to <50%. Second, although circular shaped grating blades have been proposed and used to improve the compactness of grating couplers [27], due to the use of periodic structure and light interference, integrated grating couplers still have relatively large size. Therefore, the achievable channel spacing of optical phased arrays using grating couplers is limited, and according to Eq. (5.7), it limits the beam steering range. Third, grating couplers normally are polarization dependent, and hence, using grating coupler-based optical phased array becomes problematic in practical applications, where the input signal generally has arbitrary polarization. Although the polarization diversity technique described in Section 5.2.2 can be used to support arbitrary signal, the number of components needed and the size of integrated phased array are doubled, leading to complexity and cost issues. Another major limitation of silicon integrated optical phased array using grating couplers is the limited operation bandwidth, mainly due to the periodic structure of grating couplers. This problem becomes more complicated when the wavelength tuning mechanism is used to steer the generated optical beam in one direction.

On the other hand, edge couplers-based silicon integrated optical phased arrays can achieve high signal power efficiency, broad operation bandwidth and small channel spacing. In addition, the edge coupler also supports both TE and TM polarizations, and hence, can be used for the beam steering with arbitrary input signal. Therefore, in the experimental demonstration to be discussed in the next section, where a high-speed indoor OWC system with silicon photonic integrated circuit is demonstrated, we select to use edge couplers-based integrated optical phased array for the beam steering function.

5.5 Demonstration of Indoor OWC System with Silicon Photonic Integrated Circuit

As discussed earlier in this chapter, current indoor OWC systems usually rely on discrete components, which lead to relatively complicated and bulky systems. To solve this issue, the silicon photonic integration technology provides a promising platform. In this section, we present a recent demonstration of the near-infrared indoor OWC system with a silicon photonic integrated circuit [45].

Due to laser eye and skin safety regulations, the transmission power in indoor OWC systems is limited. For example, the safe transmission power is limited to <10 dBm when the 1550 nm band is used. Therefore, the beam

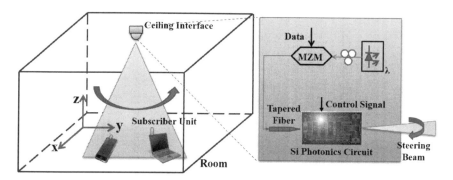

FIGURE 5.19
Structure of high-speed near-infrared indoor OWC system with a silicon photonic integrated circuit.

steering function has been proposed and used, which is unique in OWC systems. Therefore, here we focus on the use of silicon photonic integrations to realize the key beam steering function and demonstrate the high-speed data transmission using the silicon photonic integrated circuit.

The high-speed near-infrared indoor OWC system considered here is shown in Figure 5.19. In the proposed system, the silicon photonic integrated circuit is employed at the transmitter side (i.e. inside the ceiling mounted interface). Here, a tapered fiber/lensed fiber (LSF) couples the modulated signal light to the silicon integrated circuit. After passing through the integrated optical phased array, an optical beam is generated and the beam free-space propagation direction is adaptively steered towards the end user. After the OWC link, the non-imaging receiver, which consists of a CPC and a PD, is used at the subscriber side for signal collection and detection. Advanced receivers are also available, and here we just use the non-imaging receiver for simplicity.

The silicon photonic integrated circuit is an integrated optical phased array, and the structure is shown in Figure 5.20 [46]. As discussed in the previous section, due to the narrow bandwidth, high loss, polarization sensitive and relatively large channel spacing limitations of grating couplers,

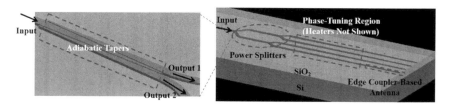

FIGURE 5.20
Structure of silicon integrated optical phased array for beam steering.

here edge couplers-based integrated optical phased array is used. The optical phased array for beam steering mainly consists of four parts, namely, the input waveguide, the power splitters, the phase shifters and edge couplers-based emitting antennas. Strip silicon waveguide is used at both the input and throughput the integrated circuit. This is due to the tight light conferment provided by strip waveguides, which reduces the signal cross-talk between neighboring waveguides. Therefore, relatively small channel spacing is allowed in the phased array and a large beam steering range becomes possible.

The power splitters are based on the adiabatic tapers introduced in Section 5.4.3. This splitter has ultra-broad operation bandwidth and <0.5 dB insertion loss. More importantly, it is compact with <10 μm length and can be operated with both TE and TM polarizations. Therefore, the size of photonic integrated circuit can be reduced, and arbitrary input signal can be supported.

To steer the beam according to practical requirement, the relative phase delay between adjacent phased array elements needs to be tuned adaptively. Either the carrier plasma dispersion effect or the thermal-optic effect can be used to introduce relative phase delay. In indoor OWC systems, since the user movement speed is limited to walking speed (<5 km/h), the beam steering speed requirement is moderate. Therefore, here we select to use the thermal-optic effect to change the relative phase delay between emitting elements. Better than millisecond tuning speed can be achieved, and it is sufficient for indoor optical OWC applications. Since the thermal-optic effect is more efficient than the carrier plasma dispersion effect, compact phase shifters can be realized.

The emitting antennas in the silicon integrated optical phased array for beam steering are based on edge couplers. An inverse taper, which is also called spot size converter (SSC) is used in the edge coupler to reduce the output waveguide width for better mode matching between the waveguide mode and the free-space mode. Through this way, high out-of-circuit power coupling efficiency and large bandwidth can be achieved.

The popular SOI wafer with 2 μm buried oxide layer and 220 nm thick silicon device layer is used as the integration platform, and here we consider a 1 × 4 integrated optical phased array. A larger number of phased array elements can be used, whilst here we just use a small dimension to demonstrate the concept. The input strip waveguide width is 500 nm, and three power splitters are used to divide the input signal into four equal parts. The length of adiabatic tapers in power splitters is 6 μm and the gap between tapers is 60 nm. After phase tuning, which is achieved by applying control signals to heaters, the width of waveguide is gradually reduced to 200 nm in the emitting antenna. As shown by Eq. (5.7), the beam steering range is mainly limited by the operation wavelength and the channel spacing of the phased array. Therefore, to increase the maximum beam steering angle, a small channel spacing is desirable. The channel spacing of the silicon integrated

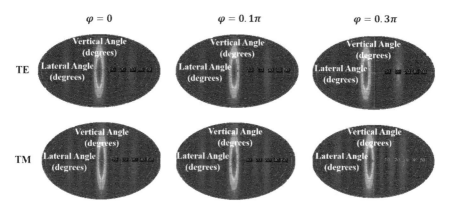

FIGURE 5.21
Simulated far-field distributions of the 1 × 4 silicon integrated optical phased array for both TE and TM polarized lights.

optical phased array is mainly limited by the crosstalk between parallel waveguides, and it is selected at 2 μm in the design, which provides a balanced trade-off between the steering range and the crosstalk.

The performance of the silicon integrated optical phased array can be studied using 3D FDTD simulations. The simulated far-field signal distributions (at 1 m away from the integrated circuit) are shown in Figure 5.21. The signal wavelength is 1550 nm with TE polarization. Similar simulations are carried out for the TM polarized light as well. The relative phase delays (φ) between adjacent phased array elements are 0, 0.1π and 0.3π, respectively. It is clear that both TE and TM polarized signals can be steered, confirming the polarization-independence of edge couplers-based silicon integrated optical phased array for indoor OWC applications. It can also be seen from the figure that side lobes also exist in the far field, and hence, it limits the beam steering range that can be achieved. The existence of side lobes is mainly because that the channel spacing selected is larger than half of the operation wavelength. Compared to typical grating couplers, the edge couplers used here have smaller size, which enables smaller channel spacing and larger steering range.

The silicon integrated optical phased array based on edge couplers is fabricated and a high-speed indoor OWC system is demonstrated with the photonic integrated circuit. The demonstration experimental setup is shown in Figure 5.22a. PRBS data is used to emulate the data to be transmitted, and it is modulated onto the optical carrier with an MZM. Then a lensed fiber with anti-reflection coating couples the light into the fabricated silicon photonic integrated circuit. As shown by the inset of Figure 5.22a, the lensed fiber is held on a high-precision positioning station, which is actively aligned to the silicon integrated circuit. After passing through the integrated optical phased array, the generated beam is launched into the free space

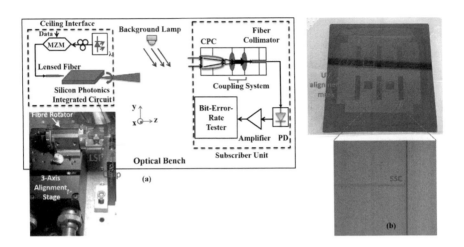

FIGURE 5.22
(a) Demonstration setup of indoor OWC system with a silicon photonic integrated circuit and
(b) photo of the fabricated circuit and the optical microscope image of SSC.

and propagates to the user side. At the receiver, a CPC is used to collect the
signal, and a fiber coupled PD with a 3 dB electrical bandwidth of about
9.6 GHz is used for optical signal detection. After detection, the converted
electrical signal is amplified and characterized using a BERT. All components in the experiment are fixed on an optical bench horizontally. To emulate the background light that always exists in indoor environments due to
sunlight or illuminations, a lamp is placed near the user side.

The silicon integrated optical phased array here provides the beam steering function. A 1 × 4 phased array is fabricated using the SOI platform.
The photonic integrated circuit is fabricated using e-beam lithography
(EBL), e-beam evaporation, inductively coupled plasma-reactive ion etching (ICP-RIE) and plasma enhanced chemical vapor deposition (PECVD).
The integrated circuit is covered with SiO_2 as the upper cladding. To achieve
smooth chip facets for efficient signal couplings, the Bosch etch process is
used. The chip facet is defined with deep-UV photolithography and etched
for about 100 μm. The photo of the fabricated circuit (with 6 chips) after the
Bosch process and the optical microscope image of the chip facet and the
SSC are shown in Figure 5.22b. In the fabricated circuit, silicon waveguides
have a dimension of 500 × 220 nm, and the width is narrowed to 200 nm in
the emitting antennas through 100 μm long SSCs. Each power splitter has
a length of 6 μm and the tip width of adiabatic tapers is 50 nm. The channel spacing between phased array elements is 2 μm to achieve a large beam
steering range and negligible crosstalk.

In the demonstration, the operation signal wavelength is 1550.12 nm, and
the power measured at the lensed fiber before the photonic integrated circuit
is 6 dBm. The total loss of silicon integrated circuit s about 4 dB and hence,

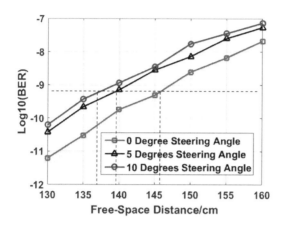

FIGURE 5.23
BER results of the indoor OWC system with a silicon photonic integrated circuit.

the power launched into the free space is about 2 dBm. Three beam steering angles are tested, which are 0°, 5° or 10°, respectively. The data to be transmitted is modulated using the simple on-off-keying (OOK) format, and the signal bit rate is 12.5 Gb/s.

After free-space propagation, the BER is measured and the results are shown in Figure 5.23. It is clear from the figure that in general, a longer free-space propagation distance lead to a worse BER. This is because of the divergence of the beam generated by the silicon integrated optical phased array, which can be expressed by Eq. (5.8). Therefore, a longer free-space distance results in a larger beam footprint and smaller collected optical power. The maximum error-free (defined as BER < 10⁻⁹) free-space distance with 0° beam steering angle is about 146 cm. When the forward-error-correction (FEC) code is applied, it is possible to achieve longer error-free OWC distances. In addition, in the demonstration no lens is used at the transmitter side. In practical applications, if a longer OWC distance is required, a collimation lens can be added at the transmitter to reduce the beam divergence and the resulting signal beam footprint.

In addition, comparing the results of systems with different beam steering angles shown in Figure 5.23, it can be seen that when the free-space propagation distance is the same, a smaller beam steering angle leads to a better BER performance. This is mainly because of the dependence of beam divergence on the steering angle. As represented by Eq. (5.8), a larger steering angle means a larger beam divergence. Therefore, a larger beam footprint is generated after the signal free-space propagation, and a worse BER is resulted.

The broadband operation advantage of the edge couplers-based silicon integrated optical phased array in the indoor OWC system is also demonstrated, by changing the operation wavelength to 1600 nm. When the bit rate

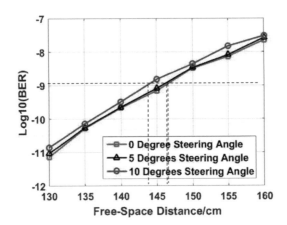

FIGURE 5.24
BER results of the indoor OWC system with a silicon photonic integrated circuit. Operation wavelength = 1600 nm.

is kept at 12.5 Gb/s, the measured BER results are shown in Figure 5.24. It can be seen that similar BER performance is achieved, and a larger steering angle also results in a worse BER performance.

In all previous measurements, the receiver at the subscriber side is placed at the center of the beam coverage area. In reality, limited mobility is desirable to allow limited user movement inside the signal coverage area. Therefore, we also characterize the BER performance while moving the subscriber unit away from the beam center. The results are shown in Figure 5.25, where the wavelength is 1550.12 nm, the beam steering angle is 0° and the OWC distance (between the transmitter and the beam center at the receiver) is either

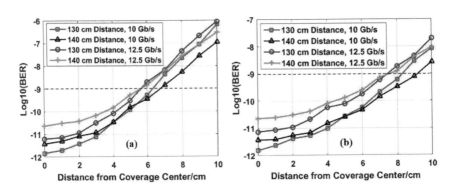

FIGURE 5.25
BER when the receiver moves inside the beam coverage area in the: (a) horizontal direction and (b) vertical direction.

130 or 140 cm. Two data rates are considered (i.e. 10 and 12.5 Gb/s). As shown by the simulated far-field in Figure 5.21, the beam intensity distribution is not circular. Therefore, we measure the BER in the horizontal and vertical directions separately (the horizontal direction is defined as the direction of optical phased array, which is the x-direction shown in Figure 5.22). Comparing the BER results in these directions, it can be seen that the BER changes more rapidly in the horizontal direction when the user moves away from the beam center. This is consistent with the far-field distribution and can also be confirmed by Eq. (5.8). Since four phased array elements is used in the horizontal direction, the beam divergence is smaller than that in the vertical direction. Therefore, the beam width is smaller after the OWC link and when moving away from the beam center, the received optical power changes more rapidly.

One key limitation of the demonstrated presented in this section is that only one-directional beam steering is achieved. In most practical applications, two-directional beam steering is normally required, which requires 2D silicon integrated optical phased arrays. With edge couplers, since the signal lights are launched into the free space along the waveguide direction from circuit facets, the vertical direction of photonic integrated circuit needs to be explored. Generally, this is challenging since photonic integration technologies only explore the lateral direction. However, in the past a few years, the multi-layer photonic integration technology that explores the vertical direction has seen rapid development [47]. The multi-layer photonic integration technology provides a promising solution to realize 2D edge couplers-based integrated phased arrays, which can enable polarization independent, low loss and broadband two-directional beam steering in indoor OWC systems.

5.6 Conclusions

Traditional indoor OWC systems rely on bulky discrete components, and in this chapter, we have discussed the recent development of using photonic integration technologies to reduce the system size, cost and weight, and to improve the system performance. We have briefly introduced the rapidly developing photonic integration technology in general and have mostly focused on the silicon photonic integration platform. Compared to other options, the silicon photonic integration technology has the key advantage of re-using existing and highly advanced CMOS electronic integrated circuit fabrication facilities. We have reviewed a number of basic building blocks on the silicon photonic integration platform, including both passive and active devices.

Compared to other fiber-based optical communication systems, the beam steering is a unique function in indoor OWC systems. Therefore, in this chapter we have focused on the beam steering device. We have reviewed the widely used MEMS-based steering mirror and the LCoS-based beam steering devices, and have analyzed their strengths and limitations. To solve the integration compatibility issue of these options, we have introduced silicon integrated optical phased array to accomplish the beam steering function in indoor OWC systems. We have presented both the fundamentals of optical phased arrays and the silicon photonic integration efforts in the past a few years.

In addition to the integrated key device and function, we have also discussed the demonstration of a high-speed indoor OWC system using the silicon photonic integrated circuit. The integrated circuit has been fabricated and tested for over 10 Gb/s data transmission through the OWC link. Results have shown the great potential of using silicon photonic integrations for future high-speed indoor OWC systems.

References

1. R. Nagarajan, C.H. Joyner, R.P. Schneider, et al., Large-scale photonic integrated circuits. *IEEE Journal of Selected Topics in Quantum Electronics*, 2005. **11**(1): pp. 50–65.
2. W. Bogaerts, D. Taillaert, B. Luyssaert, P. Dumon, J. Van Campenhout, P. Bienstman, D. Van Thourhout, R. Baets, V. Wiaux, and S. Beckx, Basic structures for photonic integrated circuits in Silicon-on-insulator. *Optics Express*, 2004. **12**(8): pp. 1583–1591.
3. R. Nagarajan, M. Kato, J. Pleumeekers, et al., InP photonic integrated circuits. *IEEE Journal of Selected Topics in Quantum Electronics*, 2010. **16**(5): pp. 1113–1125.
4. A. Boes, B. Corcoran, L. Chang, J. Bowers, and A. Mitchell, Status and potential of lithium niobate on insulator (LNOI) for photonic integrated circuits. *Laser & Photonics Reviews*, 2018. **12**(4): pp. 1700256.
5. G.T. Reed, G. Mashanovich, F.Y. Gardes, and D.J. Thomson, Silicon optical modulators. *Nature Photonics*, 2010. **4**: pp. 518–526.
6. S. Janz, Silicon-based waveguide technology for wavelength division multiplexing, in *Silicon Photonics*, L. Pavesi and D. Lockwood, Editor, Springer, Berlin, Germany, p. 323, 2004.
7. P. Dong, W. Qian, S. Liao, et al., Low loss shallow-ridge silicon waveguides. *Optics Express*, 2010. **18**(14): pp. 14474–14479.
8. Y.A. Vlasov, and S.J. McNab, Losses in single-mode silicon-on-insulator strip waveguides and bends. *Optics Express*, 2004. **12**(8): pp. 1622–1631.
9. W. Bogaerts, and S.K. Selvaraja, Compact single-mode silicon hybrid rib/strip waveguide with Adiabatic Bends. *IEEE Photonics Journal*, 2011. **3**(3): pp. 422–432.

10. D. Dai, and L. Liu, Wosinski, and S. He, Design and fabrication of ultra-small overlapped AWG demultiplexer based on alpha-Si nanowire waveguides. *Electronics Letters*, 2006. **42**(7): pp. 400–402.

11. T. Barwicz, M. Watts, M. Popovic, P. Rakich, L. Socci, F. Kartner, E. Ippen, and H. Smith, Polarization transparent microphotonic devices in the strong confinement limit. *Nature Photonics*, 2007. **1**(1): pp. 57–60.

12. H. Fukuda, K. Yamada, T. Tsuchizawa, T. Watanabe, H. Shinojima, and S. Itabashi, Ultrasmall polarization splitter based on silicon wire waveguides. *Optics Express*, 2006. **14**(25): pp. 12401–12408.

13. D. Taillaert, H. Chong, P.I. Borel, L.H. Frandsen, R M. De La Rue, and R. Baets, A compact two-dimensional grating coupler used as a polarization splitter. *IEEE Photonics Technology Letters*, 2003. **15**(9): pp. 1249–1251.

14. M. Yin, W. Yang, Y. Li, X. Wang, and H. Li, CMOS-compatible and fabrication-tolerant MMI-based polarization beam splitter. *Optics Communications*, 2015. **335**(1): pp. 48–52.

15. H. Fukuda, K. Yamada, T. Tsuchizawa, T. Watanabe, H. Shinojima, and S. Itabashi, Silicon photonic circuit with polarization diversity. *Optics Express*, 2008. **16**(7): pp. 4872–4880.

16. W.W. Lui, W.P. Huang, K. Yokoyama, and W.-P. Huang, Polarization rotation in semiconductor bending waveguides: A coupled-mode theory formulation. *Journal of Lightwave Technology*, 1998. **16**(5): pp. 929–936.

17. J. Zhang, M.B. Yu, G.Q. Lo, and D.L. Kwong, Silicon-waveguide-based mode evolution polarization rotator. *IEEE Journal of Selected Topics in Quantum Electronics*, 2010. **16**(1): pp. 53–60.

18. Bowers, D.D., and J.E. Bowers, Novel concept for ultracompact polarization splitter-rotator based on silicon nanowires. *Optics Express*, 2011. **19**(11): pp. 10940–10949.

19. H. Yu, M. Chen., P. Li, S. Yang, H. Chen, and S. Xie, Silicon-on-insulator narrow-passband filter based on cascaded MZIs incorporating enhanced FSR for down-converting analog photonic links. *Optics Express*, 2013. **21**(6): pp. 6749–6755.

20. W. Bogaerts, P. Dumon, D. Thourhout, et al., Compact wavelength-selective functions in silicon-on-insulator photonic wires. *IEEE Journal of Selected Topics in Quantum Electronics*, 2006. **12**(6): pp. 1394–1401.

21. P. Dong, N.-N. Feng, D. Feng, et al., GHz-bandwidth optical filters based on high-order silicon ring resonators. *Optics Express*, 2010. **18**(23): pp. 23784–23789.

22. G. Jiang, R. Chen, Q. Zhou, et al., Slab-modulated sidewall Bragg gratings in silicon-on-insulator ridge waveguides. *IEEE Photonics Technology Letters*, 2011. **23**(1): pp. 6–9.

23. K. Sasaki, F. Ohno, A. Motegi, and T. Baba, Arrayed waveguide grating of $70 \times 60 \ \mu m^2$ size based on Si photonic wire waveguides. *Electronics Letters*, 2005. **41**(14): pp. 801–802.

24. Yariv, A., Coupled-mode theory for guided-wave optics. *IEEE Journal of Quantum Electronics*, 1973. **9**(9): pp. 919–933.

25. Y. Zhang, S. Yang, A. Lim, G. Lo, C. Gal-land, T. BaehrJones, and M. Hochberg, A compact and low loss Y-junction for submicron silicon waveguide. *Optics Express*, 2013. **21**(1): pp. 1310–1316.

26. Z. Sheng, Z. Wang, C. Qiu, et al., A compact and low-loss MMI coupler fabricated with CMOS technology. *IEEE Photonics Journal*, 2012. **4**(6): pp. 2272–2277.

27. S. Gao, Y. Wang, K. Wang, and E. Skafidas, High contrast circular Bragg grating reflector on silicon-on-insulator platform. *Optics Letters*, 2016. **41**(3): pp. 520–523.
28. W. Shi, X. Wang, C. Lin, et al., Silicon photonic grating-assisted, contra-directional couplers. *Optics Express*, 2013. **21**(3): pp. 3633–3650.
29. Hochberg, L.C., and M. Hochberg, *Silicon Photonics Design—From Devices to Systems*. Cambridge, UK: Cambridge University Press, 2015.
30. M. Nedeljkovic, R. Soref, and G.Z. Mashanovich, Free-carrier electro-refraction and electroabsorption modulation predictions for silicon over the 1–14 micron infrared wavelength range. *IEEE Photonics Journal*, 2011. **3**(6): pp. 1171–1180.
31. D.J. Thomson, F.Y. Gardes, Y. Hu, G. Mashanovich, M. Fournier, P. Grosse, J.-M. Fedeli, and G.T. Reed, High contrast 40Gbit/s optical modulation in silicon. *Optics Express*, 2011. **19**(12): pp. 11507–11516.
32. P.S. Reano, and P. Sun, Submilliwatt thermo-optic switches using free-standing silicon-on-insulator strip waveguides. *Optics Express*, 2010. **18**(8): pp. 8406–8411.
33. T. Baba, S. Akiyama, M. Imai, N. Hirayama, H. Takahashi, Y. Noguchi, T. Horikawa, and T. Usuki, 50-Gb/s ring-resonator-based silicon modulator. *Optics Express*, 2013. **21**(10): pp. 11869–11876.
34. J.D. Bradley, P.E. Jessop, and A.P. Knights, Silicon waveguide-integrated optical power monitor with enhanced sensitivity at 1550 nm. *Applied Physics Letters*, 2005. **86**(24): pp. 241103.
35. C.T. DeRose, D.C. Trotter, W.A. Zortman, A.L. Starbuck, M. Fisher, M.R. Watts, and P.S. Davis, Ultra compact 45 GHz CMOS compatible Germanium waveguide photodiode with low dark current. *Optics Express*, 2011. **19**(25): pp. 24897–24904.
36. V.A. Aksyuk, F. Pardo, D. Carr, et al., Beam-steering micromirrors for large optical cross-connects. *Journal of Lightwave Technology*, 2003. **21**(3): pp. 634–642.
37. K.H. Koh, T. Kobayashi, and C. Lee, A 2-D MEMS scanning mirror based on dynamic mixed mode excitation of a piezoelectric PZT thin film S-shaped actuator. *Optics Express*, 2011. **19**(15): pp. 13812–13824.
38. K. Wang, A. Nirmalathas, C. Lim, and E. Skafidas, High-speed optical wireless communication system for indoor applications. *IEEE Photonics Technology Letters*, 2011. **23**(8): pp. 519–521.
39. P.F. McManamon, T.A. Dorschner, D.L. Corkum, et al., Optical phased array technology. *Proceedings of IEEE*, 1996. **84**(2): pp. 268–298.
40. Y. Wang, S. Gao, K. Wang, and E. Skafidas, Ultra-broadband and low-loss 3 dB optical power splitter based on adiabatic tapered silicon waveguides. *Optics Letters*, 2016. **41**(9): pp. 2053–2056.
41. S. Yang, M.L. Cooper, P.R. Bandaru, and S. Mookherjea, Giant birefringence in multi-slotted silicon nanophotonic waveguides. *Optics Express*, 2008. **16**(11): p. 8306–8316.
42. M. Pu, L. Liu, W. Xue, et al., Tunable microwave phase shifter based on silicon-on-insulator microring resonator. *IEEE Photonics Technology Letters*, 2010. **22**(12): pp. 869–871.
43. J. Sun, E. Timurdogan, A. Yaacobi, E.S. Hosseini, and M.R. Watts, Large-scale nanophotonic phased array. *Nature*, 2013. **493**(1): pp. 195–199.
44. J.C. Hulme, J.K. Doylend., M.J.R. Heck, J.D. Peters, M.L. Davenport, J.T. Bovington, L.A. Coldren, and J.E. Bowers, Fully integrated hybrid silicon two dimensional beam scanner. *Optics Express*, 2015. **23**(5): pp. 5861–5874.

45. K. Wang, A. Nirmalathas, C. Lim, E. Wong, K. Alameh, H. Li, and E. Skafidas, High-speed indoor optical wireless communication system employing a silicon integrated photonic circuit. *Optics Letters*, 2018. **43**(13): pp. 3132–3135.
46. K. Wang, Z. Yuan, E. Wong, K. Alameh, H. Li, K. Sithamparanathan, and E. Skafidas, Experimental demonstration of indoor infrared optical wireless communications with a silicon photonic integrated circuit. *Journal of Lightwave Technology*, 2019. **37**(2): pp. 619–626.
47. W. D. Sacher, Y. Huang, G.-Q. Lo, and J. Poon, Multilayer silicon nitride-on-silicon integrated photonic platforms and devices. *Journal of Lightwave Technology*, 2015. **33**(4): pp. 901–910.

6

Indoor Optical Wireless Localization Technology

Due to the large license-free bandwidth advantage, the optical wireless technology is an attractive option to provide ultra-high-speed wireless communications in indoor environments. In previous chapters, we have discussed some demonstrations of high-speed indoor OWC systems using the near-infrared wavelength range. Due to the safety regulation, the maximum transmission power in indoor OWC systems is limited. Considering the aperture size limit of the receiver, only limited signal coverage inside the room can be provided. Therefore, in the example presented in Section 2.3, the combination of limited mobility and indoor user localization function is utilized.

The localization function provides the location information of users, and hence, it is capable of tracking the movement of users. Therefore, in addition to high-speed wireless data transmissions, the indoor user localization is also highly demanded in a large number of promising applications, such as the location-based advertisement in shopping centers and the location-based smart elderly care.

In this chapter, we will firstly provide an overview of current indoor user localization techniques in Section 6.1. Then we will introduce the basic principle of indoor localization using the optical wireless technology in Section 6.2. We will discuss the localization accuracy limiting factors and corresponding improvement techniques in Section 6.3. Since the height information of users is also highly desirable to determine the beam steering angle in indoor OWC systems, we will describe the 3D indoor optical wireless localization scheme in Section 6.4. Finally, we will summarize this chapter in Section 6.5.

6.1 Indoor Localization Systems

As discussed above, the indoor localization function is highly demanded in indoor OWC systems to solve the signal transmission power limit set by laser eye and skin safety requirement. The indoor localization system provides the location information of users and keep tracking the movement of users to control the beam steering device, such as the MEMS mirror or the integrated optical phased array discussed in Chapter 5. In this section, we will briefly overview current indoor localization techniques.

GPS has been widely used for outdoor localizations and navigation. The GPS localization system relies on GPS satellites high above the Earth, which send out localization signals. When sending out the localization message, the time that the message is transmitted and the location of satellite when the message is sent are included. A GPS receiver then calculates its location by precisely timing the signals sent by GPS satellites. Therefore, the GPS technique is based on the signal time-of-arrival information. The typical localization accuracy of GPS systems is in the range of several meters. With more advanced accuracy enhancement techniques, such as carrier phase tracking, the precision can be improved to be better than several tens of centimeters.

However, the use of GPS technology in indoor applications is practically limited. This is mainly because that the GPS signals sent out by satellites suffer from significant attenuation due to physical shadowing in indoor environments, which lead to frequent loss of localization signals. In addition, GPS signals also experience severe multipath dispersion, and hence, the localization accuracy is substantially degraded in indoor environments.

To solve these limitations and to realize accurate indoor localizations, several technologies have been proposed and investigated. The most widely studied and used option is the RF signal-based technology. Various RF signals and standards have been explored, such as the Wi-Fi signal using either the 2.4 GHz or the 5 GHz band, the Bluetooth signal at the 2.4 GHz band, the UWB signal and the radio-frequency identification (RFID)-based scheme [1–7]. An example of RF-based indoor localization system is shown in Figure 6.1 (a), where multiple RF transmitters are utilized. The RF transmitters are used for both communications and localizations. The user localization function can be realized exploring different properties the received signals, such as the received signal strength, the signal angle-of-arrival and

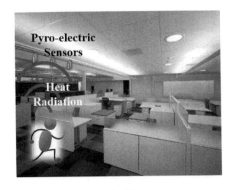

(a)   (b)

FIGURE 6.1
Principle of indoor user localizations. (a) RF-based localization system and (b) pyro-electric sensor-based localization system.

the signal time-of-arrival information. Take the received signal strength-based scheme as an example. Since the RF signal attenuation is proportional to the wireless transmission distance, the distance between the RF transmitter and the receiver located at the user side can be estimated by measuring the received signal strength [1]. With signals from multiple transmitters, the location of user can then be estimated. This type of localization principle is normally referred as the non-map-based scheme, since no signal strength map is pre-measured. The map-based scheme is also widely used, where a number of reference points are pre-defined in the indoor environment, and the RF signal strengths at these reference points are pre-measured. The received signal strengths are then compared to those of reference points to estimate the user location. Typically, the map-based solution can achieve better localization accuracy than the non-map-based, with the cost of pre-measuring reference points. Both map-based and non-map-based principles can be used in the RF localization systems exploring signal angle-of-arrival and time-of-arrival information as well. The major limiting factors in RF-based indoor localization systems are the signal attenuation, back reflection and multipath dispersion, and the typical localization accuracy in indoor personal living/working spaces is several tens of centimeters to about one meter.

In addition to RF signal-based indoor localization systems, the infrared sensors have also been explored to provide the user location in indoor environments. One example is the pyro-electric sensor-based indoor localization principle [8], as shown in Figure 6.1b. Pyro-electric sensors are widely deployed in auto-doors and they are capable of monitoring the change of heat radiation received. When the user moves inside the room, the heat radiation detected by the pyro-electric sensors changes accordingly, and by combining the changes observed by multiple sensors, the location of user can be estimated. The localization accuracy is better than most RF signal-based schemes, and the typical localization error is within half a meter. However, pyro-electric sensor-based localization scheme usually can only locate moving users, and the performance can be easily affected by environmental changes.

In addition, the imaging sensors also have the capability of providing the indoor localization function [9]. The user location can be estimated by various image processing techniques, and 3D user location with height information can be obtained. The rapidly developing neural networks have also been applied in processing the images captured by imaging sensors, and improved localization accuracy can be achieved [10]. The imaging sensor can also be combined with LEDs to achieve the indoor localization function [11]. For example, by arranging multiple LEDs into a special pattern and processing the LED pattern captured by the imaging sensor, indoor localization of the receiver can be realized. The typical localization accuracy of imaging sensors-based indoor localization systems is in the order of several centimeters to tens of centimeters. However, the image processing requires high computation costs, and the imaging sensor in personal areas may invade personal privacy.

Another demonstrated indoor localization scheme is using passive infrared beams [12]. In this scheme, multiple infrared transmitters are installed in the corners of an indoor environment (e.g. office) and a receiver is held by the user. Each transmitter sends out the localization beam to a certain direction, and both the location and the beam transmission direction of each transmitter is pre-known. The location information of the user is then estimated by calculating the interception area of received beams from different transmitters. In this localization scheme, the accuracy depends on the transmitter configurations, including the number of transmitters installed, the locations of transmitters, the radiation directions and the beam divergences. An accuracy of about 0.6 m has been experimentally demonstrated with three transmitters [12].

6.2 Indoor Localization Based on the Optical Wireless Technology – "Search and Scan"

As described in detail in Section 6.1, a number of indoor localization technologies have already been proposed and demonstrated. However, they have several limitations of the application in high-speed near-infrared indoor OWC systems. The most important limitation is that they require additional devices in addition to the OWC ones just for the localization purpose, which lead to increased system complexity and cost (as well as additional power consumption). In addition, the localization accuracy that can be achieved is also limited. Due to the highly limited transmission power in indoor OWC systems, accurate user location information is needed to ensure correct wireless data transmissions. Therefore, the use of some previously developed indoor localization schemes may lead to incorrect signal coverage area and service interruptions. To solve these issues, the re-use of available devices in indoor OWC systems to fulfill the indoor localization function has been proposed and investigated [13]. We will focus on this optical wireless technology-based indoor user localization principle in this section.

In high-speed near-infrared indoor OWC systems, such as the example discussed in Section 2.3, the beam steering device is normally utilized. This beam steering capability can be explored for the indoor localization function, by using the "search and scan" principle. Here we use a simple case to explain this localization principle, where there is one OWC transceiver located at the ceiling for high-speed data transmissions, and each user is equipped with an OWC user transceiver, as shown in Figure 6.2a. A beam steering device is used inside the ceiling mounted OWC transceiver to control the free-space propagation direction of the signal beam. For localization and data transmission purposes, we divide the room to be covered into cells, and the optical beam covers one cell at a time to enable limited mobility within the cell.

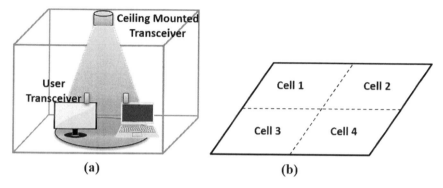

FIGURE 6.2

Indoor optical wireless-based localization system principle. (a) Typical indoor optical wireless system structure and (b) the division of room space into cells.

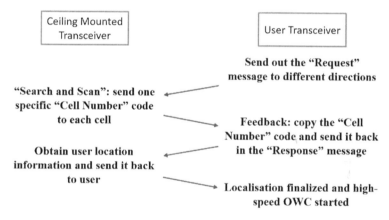

FIGURE 6.3

The principle of "search and scan"-based indoor optical wireless localization.

For simplicity, we assume that there are four equal-sized cells inside the room, as shown in Figure 6.2b. Each cell is assigned with a unique "cell number".

The "search and scan"-based indoor optical wireless localization principle is illustrated by Figure 6.3. The localization function is initiated by the user, who sends out a "Request" message to the ceiling mounted OWC transceiver. The optical wireless signal transmission from the user to the ceiling mounted interface, which is normally referred as the uplink, has already been developed using the similar principle as the downlink (i.e. from the ceiling interface to the user), and more details about the uplink can be found in reference [14]. Since the user does not know the location of the ceiling mounted interface, the "Request" message is sent out to difference directions to ensure the full coverage over the room ceiling. Upon receiving the "Request" message, the ceiling mounted

OWC transceiver starts the "Search and Scan" process by sending out the specific "Cell Number" code to each cell. The cell size is determined by the OWC speed required, and a smaller cell is used when a higher speed is required.

After free-space transmission, the "Search and Scan" message arrives at the user side. The user transceiver copies the unique "Cell Number" code and sends it back in the "Feedback" message. From the received "Feedback" message, the ceiling mounted OWC transceiver is able to obtain the location information of the user (i.e. which cell the user is located). Then the ceiling mounted transceiver sends this location information back to the user. At this time both the ceiling mounted transceiver and the user transceiver have the location information and hence, high-speed OWC link can be established for data transmissions. This "Search and Scan" process continues to update the user's localization information, to cope with the case where the user moves out of the initially located cell.

The feasibility of the "Search and Scan" principle-based indoor optical wireless localization scheme discussed above has been experimentally demonstrated, where a MEMS mirror is used to provide the beam steering and scanning capability [13]. Results have shown that the user location (i.e. the cell that the user is located) can always be obtained correctly through the unique "Cell Number" code.

6.3 Indoor Optical Wireless Localization Principles with High Accuracy

From the working principle of the "Search and Scan"-based indoor optical wireless localization scheme described in the previous section, it can be seen that the localization accuracy is equal to the size of cell divided inside the indoor environment. Therefore, if more accurate location information is needed, a small cell size is needed, and a large number of cells need to be searched and scanned. In consequence, this "Search and Scan" process takes a long time to complete. Since the data transmission via OWC link cannot be started until the user localization information is obtained, the overall OWC system throughput is significantly reduced.

To overcome this limitation and to achieve better localization accuracy, the received signal strength information can be utilized [15]. This is based on the fact that the optical intensity distribution of the localization signal at the user end is approximately Gaussian, since the laser is used as the light source in the system, the direct LOS channel is used and the free-space transmission distance is only several meters in indoor applications. Therefore, the field distribution profile of the transmitted light signal is mostly preserved. In practical cases, the beam footprint of the localization beam is much larger than receiver aperture at the user side. Therefore, we assume that the power is evenly distributed over the entire receiver aperture, and the received localization signal power P_r can be expressed as:

$$P_r = \frac{2P_tS_r}{\pi\omega^2} exp\left(-\frac{2r^2}{\omega^2}\right) \tag{6.1}$$

where S_r is the input aperture of the receiver at the user side, P_t is the transmission power of the "Search and Scan" localization signal, ω is the beam footprint of the localization beam at the receiver side and r is the distance from center of localization beam to the user location. Here we assume that the user receiver is placed on a reception plane with a fixed height. Based on Eq. (6.1), if the transmission power from the ceiling mounted transceiver and the beam footprint after free-space propagation are included in the "Search and Scan" message sent out by the ceiling transceiver, since the receiver aperture size is pre-known to the user, the distance from the localization beam center to the user on the reception plane can be estimated by measuring the received signal strength.

To obtain the exact coordinates (x and y) of user, the ceiling mounted OWC transmitter just needs to steer the localization beam in both x- and y-directions by half of the cell size and sends out the "Search and Scan" message with known power and beam footprint again. The distance from user to the new localization beam center can be estimated according to Eq. (6.1). By combing the r values estimated and the beam center information that is already known, the exact coordinate of user can be estimated. Through this way, a localization accuracy better than the localization beam footprint (or equivalently, the cell size) can be achieved.

The entire localization process of the received signal strength-based principle is as follows:

1. Divide the room into several cells and assign each cell with a specific "Cell Number" code. The size of the cell has some impact on the achievable localization accuracy, and it will be analyzed later.

2. The user activates the localization function by sending out a "Request" message. The message is sent to different directions by the user transceiver, since the locations of user and ceiling mounted OWC transceiver are unknown;

3. After receiving the "Request" message, the ceiling mounted OWC transceiver starts the "Search and Scan" process by sending out the specific "Cell Number" code to each cell. The cell scanning is realized by the beam steering device. To enable better than cell size localization accuracy through exploring the received signal strength, the transmission power and beam footprint information of the localization beam are added to the "Search and Scan" message;

4. After receiving the "Search and Scan" message, the user measures the received signal power P_r. The user then uses Eq. (6.1) together with the known receiver aperture size S_r, transmission power P_t and beam footprint ω, to estimate the distance r from the beam center. Afterwards the user copies the "Cell Number" information and sends out the "Response" message to the ceiling mounted OWC transceiver. The estimated distance r between the localization beam center and the user on the reception plane is included;

5. When receiving the "Response" message, the ceiling mounted OWC transceiver steers the localization beam in the x-direction by half of the cell size and repeats steps 2 and 3;

6. The ceiling mounted OWC transceiver then steers the localization beam in the y-direction by half of the cell size the repeat steps 2 and 3;

7. Using the three estimated r information and the corresponding beam center information, both the ceiling mounted interface and the user transceiver then estimate the exact coordinates of the user in the room;

8. With the accurate user location information, establish high-speed data transmissions via the indoor OWC link.

Based on the localization procedure described above, accurate location of the user can be achieved. Here we present a demonstration of the received signal strength-based accurate indoor optical wireless localization principle. In the experiment, we use the 1550 nm wavelength for the localization functions, and set the localization beam footprint at 2 m. The bit rate of the localization messages, such as the "Search and Scan" message, is 50 Mbps with the OOK modulation format. Different from data transmissions, the data speed requirement in the localization function is low to moderate, and hence, we select a relatively low data rate to maximize the receiver sensitivity.

When the transmission power of the localization signal is 7 mW, which is within the laser safety regulation, the estimated user locations are shown in Figure 6.4. The user transceiver is moved along the $y = 2$ m line. It is clear from the results that the localization of user can always be achieved using the method we discussed above, and the average localization accuracy is about 15.26 cm. Compared to the localization beam footprint and the cell size, much better accuracy is achieved using the received signal strength information.

We can also further analyze the causes of the remaining localization error. The localization error is mainly due to the impact of background light. In the experiment, strong background light from illumination lamps are included. The impact of background light can be analyzed based on Eq. (6.1), where the received power P_r measured in real experiment actually consists of two parts (i.e. the localization signal power and the background light power). Therefore, the received power P_r that is measured by the user transceiver is always higher than the actual localization signal power and hence the

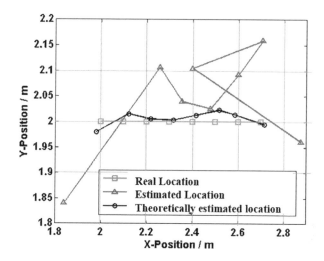

FIGURE 6.4
Localization results using the received signal strength-based indoor optical wireless localization principle.

localization accuracy is deteriorated. Using Eq. (6.1), the error on the estimation of distance from beam center r due to the existence of background light, which is denoted as r_{error}, can be expressed as:

$$r_{error} = \sqrt{-\frac{\omega^2}{2}\ln\frac{\pi\omega^2 P_{rs}}{2P_t S_r}} - \sqrt{-\frac{\omega^2}{2}\ln\frac{\pi\omega^2 (P_{rs}+P_{rb})}{2P_t S_r}} \qquad (6.2)$$

where P_{rs} is the received localization signal power and P_{rb} is the received background light power.

The localization error due to background light is also experimentally investigated. Here we switch off all the illumination lamps and insert an optical bandpass filter with a narrow 0.6 nm full-width half-maximum (FWHM) passband to filter out all out-of-band background light. The remaining background light collected by the user transceiver is negligible. Then we repeat the measurements and the localization process. The results are shown by the black line in Figure 6.4. It can be seen that without the impact of background light, the average localization error is only about 2.25 cm and the maximum localization error is measured to be about 2.83 cm. Therefore, much better localization accuracy is achieved. Measurement results also confirm that in practice, the background light limits the achievable localization accuracy in the received signal strength-based indoor localization principle.

In addition to the background light, there are two other factors leading to the localization error. The first one is that the actual field distribution is not ideally Gaussian at the receiver side, due to the imperfections of devices and the free-space signal propagation. The other reason is that the power is not evenly distributed over the receiver aperture, especially when the receiver size increases. Therefore, the use of Eq. (6.1) for distance estimation has systematic errors. However, as shown in Figure 6.4, these two factors only lead to moderate localization error.

From Eq. (6.2), it can also be seen that the localization accuracy depends on the localization signal transmission power as well. To investigate the impact, we change the localization beam transmission power from 1 mW to 3 mW and estimate the corresponding user location using the received signal strength-based principle. The average localization error is shown in Figure 6.5. The localization beam footprint is fixed at 1 m and the user transceiver moves along the $y = 2$ m line. It can be seen from the figure that the average localization error decreases with the increasing transmission power. This is because that with a larger transmission power, the received localization signal power increases, whilst the background light power remains constant. Therefore, according to Eq. (6.2), the impact of background light is reduced.

From the discussions above, it is clear that the major limit in the received signal strength-based indoor optical wireless localization principle is the background light. To further improve the localization accuracy and to provide better robustness against background light, here we introduce a near-infrared indoor optical wireless localization principle with the background light power estimation capability. This principle is also based on the "Search

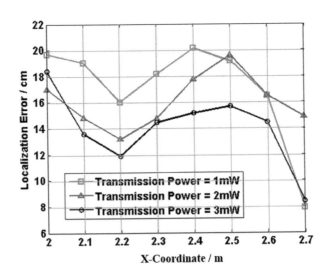

FIGURE 6.5
Localization results using the received signal strength-based indoor optical wireless localization principle with different transmission powers.

and Scan" process described above and also explores the received signal strength information. To mitigate the localization accuracy degradation due to background light, here two localization signals are transmitted. These localization signal beams have different transmission power levels, P_{t1} and P_{t2}, respectively, and they are transmitted at adjacent time instances. Due to the slow dynamics feature in typical indoor environments, where the moving speed of user is generally low (e.g. walking speed) and the environmental changes are slow and infrequent, the background light power collected during these two adjacent time instances can be considered as approximately constant. The user location can also be considered as unchanged. Therefore, using Eq. (6.1), the received signal strengths according to the two localization signals, which are denoted as P_{rs1} and P_{rs2}, respectively, can be expressed as:

$$P_{rs1} = P_{r1} - P_{rb} = \frac{2P_{t1}S_r}{\pi\omega^2}exp\left(-\frac{2r^2}{\omega^2}\right)$$ (6.3)

$$P_{rs2} = P_{r2} - P_{rb} = \frac{2P_{t2}S_r}{\pi\omega^2}exp\left(-\frac{2r^2}{\omega^2}\right)$$ (6.4)

where P_{r1} and P_{r2} are the measured received optical powers during the two adjacent time instances, respectively. The beam footprint is kept unchanged for these two localization signals. Using Eq. (6.3) and Eq. (6.4), the received background light power can be calculated as:

$$P_{rb} = \frac{P_{r2}P_{t1} - P_{r1}P_{t2}}{P_{t1} - P_{t2}}$$ (6.5)

With the background light power P_{rb} being estimated, the localization accuracy can be improved significantly. The distance between the user location and the localization beam center r on the reception plane can now be calculated as:

$$r^2 = -\frac{\omega^2}{2}ln\left(\frac{P_{r1} - P_{r2}}{P_{t1} - P_{t2}} \cdot \frac{\pi\omega^2}{2S_r}\right)$$ (6.6)

By steering the localization beam along both x- and y-directions by half of the beam footprint and performing the same measurements and estimations described above, the exact location (i.e. coordinates information) of the user can be obtained. Compared to the basic received signal strength-based method, results show that with the background light estimation principle, the average localization error can be significantly reduced [16].

6.4 Three-Dimensional Indoor Optical Wireless Localization

In the optical wireless technology-based indoor localization schemes discussed in the previous section, it is always assumed that the user transceiver moves on the reception plane, which has a fixed height. The height of the reception plane and the height difference between the ceiling mounted OWC transceiver and the user transceiver are also assumed to be pre-known. These assumptions are required to fix the localization beam footprint after free-space propagation. However, in real applications, the height of the user transceiver can be changing and the height difference between the localization transmitter and receiver is unpredictable. Therefore, it is high desirable to have the 3D localization capability, where the height of the user transceiver can be estimated as well. In this section, we introduce the 3D indoor optical wireless localization principle, which explores both the received signal strength and the signal angle-of-arrival information simultaneously to enable the height estimation capability [17].

In the 3D indoor optical wireless localization system, instead of using the simple non-imaging receiver, the more advanced single-channel imaging receiver needs to be used to provide the signal angle-of-arrival information [18,19]. The basic structure of single-channel imaging receiver is shown in Figure 6.6. The single-channel imaging receiver mainly consists of an imaging lens for signal collection and focusing, and a PD attached to a 2-axis actuator. The PD is placed on the back focal plane of the imaging lens, which has a focal length of f. The actuator is based on the voice-coil, which has been widely used in hard disks and CD read heads. By changing the current passing through the voice-coil, the magnetic field generated can move the actuator arm and the attached PD precisely on the focal plane.

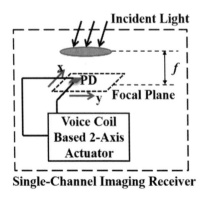

FIGURE 6.6
Structure of the single-channel imaging receiver.

In the indoor OWC and localization system, since the optical signal beam footprint commonly is much larger than the aperture of the imaging lens, all signal lights incident onto the lens can be approximately seen as parallel. Therefore, the incident lights are focused onto a small spot on the focal plane of the lens in the single-channel imaging receiver. Then the 2-axis actuator moves the PD on the focal plane to search for the focused signal spot. If the photocurrent from the PD increases sharply over a threshold when the PD is moved across the focal plane, the focused light spot is found [19]. Since the PD used in the single-channel imaging receiver has a relatively small size (generally tens of micrometers to a few hundred micrometers), only the background light that has a similar incident angle as the signal light is collected and detected. Therefore, the single-channel imaging receiver can reject most of the background light that always exists in indoor environments due to both sunlight and illumination lamps.

In addition to rejecting most background light, the single-channel imaging receiver has another capability, which is providing the signal angle-of-arrival information. In the single-channel imaging receiver, the location of light spot on the focal plane of the imaging lens is known, which is the same location as the PD attached to the 2-axis actuator. Therefore, when combined with the pre-known focal length of imaging lens f, as shown in Figure 6.7, the signal angle-of-arrival, which is denoted as φ, can be calculated. Here we assume the distances between the focused light spot and the center of the imaging lens on the focal plane in x- and y-directions (as defined in the figure) are d_{xsl} and d_{ysl}, respectively, which can be obtained from the PD location. Then the signal angle-of-arrival in x- and y-directions (φ_x, φ_y) can be calculated as:

$$\tan\left(\varphi_x\right) = d_{xsl} / f \tag{6.7}$$

$$\tan\left(\varphi_y\right) = d_{ysl} / f \tag{6.8}$$

The signal angle-of-arrival information, together with the measured received signal strength information, can be jointly explored to realize the 3D indoor localization function with height estimation. In the coordinate shown in Figure 6.7, here we assume that the location of ceiling mounted OWC transceiver is (0, 0, 0). For a pre-defined height difference h_0 in the z-direction between the ceiling mounted OWC transceiver and the user transceiver, the location of beam center after free-space propagation $\left(x_{b0}, y_{b0}, h_0\right)$ and the beam footprint ω_0 can be known, since they are directly related with the beam orientation information and the beam divergence information, respectively. These are added into the "Search and Scan" message and transmitted to the user side. Assume the real height difference in the z-direction between the ceiling mounted OWC transceiver and the user transceiver is h_{diff}, then

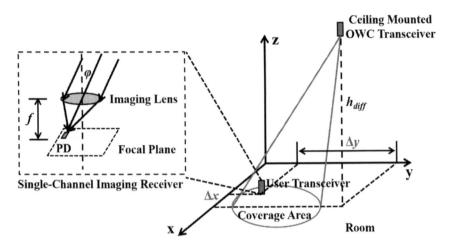

FIGURE 6.7
Signal angle-of-arrival information estimation principle.

the real beam footprint ω at the user side can be calculated using the geometric optics approximation, which can be expressed as:

$$\omega = \frac{h_{diff}}{h_0} \cdot \omega_0 \tag{6.9}$$

Accordingly, the real beam center coordinate at the user side (x_b, y_b) can be described as:

$$x_b = \frac{h_{diff}}{h_0} \cdot x_{b0} \tag{6.10}$$

$$y_b = \frac{h_{diff}}{h_0} \cdot y_{b0} \tag{6.11}$$

Since the signal incident angle information (φ_x, φ_y) is already obtained via the location of PD on the focal plane using Eq. (6.7) and Eq. (6.8), the location of the user transceiver (x_u, y_u, h_{diff}) can be expressed as:

$$x_u = h_{diff} \cdot \tan(\varphi_x) \tag{6.12}$$

$$y_u = h_{diff} \cdot \tan(\varphi_y) \tag{6.13}$$

Therefore, the distance between the user transceiver and the localization beam center in the x- and y-directions on the reception plane with height h_{diff}, which are denoted as d_{xub} and d_{yub}, and the Euclidean distance r_{ub}, can be calculated as:

$$d_{xub} = x_b - x_u = \frac{h_{diff}}{h_0} \cdot x_{b0} - h_{diff} \cdot \tan(\varphi_x) \quad (6.14)$$

$$d_{yub} = y_b - y_u = \frac{h_{diff}}{h_0} \cdot y_{b0} - h_{diff} \cdot \tan(\varphi_y) \quad (6.15)$$

$$r_{ub} = \sqrt{d_{xub}^2 + d_{yub}^2} \quad (6.16)$$

At the single-channel imaging receiver side, the received optical power P_r can be measured. Assume the localization signal transmission power is P_t. Since the optical intensity distribution is still approximately Gaussian, the received power can be expressed by Eq. (6.1). Combining the received signal strength information described by Eq. (6.1) and the received signal angle-of-arrival information described by Eq. (6.7) to Eq. (6.16), the real height different between the ceiling mounted fiber transceiver and the user transceiver h_{diff} and the real beam footprint at the user side ω can be calculated as:

$$h_{diff} = \frac{h_0}{\omega_0} \cdot \sqrt{\frac{2P_tS_r}{\pi P_r} \cdot \exp\left(-2h_0^2 \cdot \frac{\left(\frac{x_{b0}}{h_0} - \frac{d_{xsl}}{f}\right)^2 + \left(\frac{y_{b0}}{h_0} - \frac{d_{ysl}}{f}\right)^2}{\omega_0^2}\right)} \quad (6.17)$$

$$\omega = \frac{2P_t \cdot \sqrt{S_r}}{\pi P_r} \cdot \exp\left(-2h_0^2 \cdot \frac{\left(\frac{x_{b0}}{h_0} - \frac{d_{xsl}}{f}\right)^2 + \left(\frac{y_{b0}}{h_0} - \frac{d_{ysl}}{f}\right)^2}{\omega_0^2}\right) \quad (6.18)$$

Using Eqs. (6.17) and (6.18), measuring the received signal strength, and obtaining the location of PD on the focal plane in the single-channel imaging receiver, both the real height of the user transceiver and the location on the reception plane can be estimated. Therefore, 3D indoor localization is achieved. This principle has been demonstrated. An average height estimation accuracy of better than 6 cm and an average localization accuracy on the reception plane of better than 7 cm are achieved simultaneously [17].

6.5 Conclusions

Due to the highly limited signal transmission power in high-speed near-infrared indoor OWC systems, which is limited by the laser safety regulation, the indoor user localization function is highly demanded. Therefore, we have focused on obtaining the user location information in indoor environments in this chapter.

We have firstly briefly reviewed currently available indoor localization principles in Section 6.1. It has been shown that due to the high signal attenuation, shadowing and blockage, the GPS technology, which is widely used for outdoor navigations, is not suitable for indoor applications. Therefore, a number of alternatives have been proposed and developed. In general, the RF-based technologies using various frequency bands and standards are limited by the signal reflection, diffraction and multipath dispersion, which result in relatively low localization accuracy; the pyro-electric sensor-based localization scheme is vulnerable to environmental changes and faces challenges to detect static users; the image sensor-based principle may invade personal privacy and has high computation cost; and the passive infrared beams-based system is highly dependent on the number of transceivers used and the transceiver configuration. In addition, these developed technologies require additional dedicated devices just for the localization purpose when used in indoor OWC systems.

To overcome these limitations, we have introduced the optical wireless-based indoor localization principle in Section 6.2. The basic localization principle is based on the "Search and Scan" process, where the indoor environment is firstly divided into equal-sized cells and the ceiling mounted OWC transceiver sends the unique "Cell Number" code to each cell and scans the room. Upon receiving the unique code, the user transceiver copies it and sends it back. From this feedback message, the location information (i.e. which cell the user is located) is obtained. The simple optical wireless-based indoor localization principle does not require any additional hardware, and hence, can be realized with low cost and complexity.

In the "Search and Scan"-based indoor optical wireless localization principle, the localization accuracy equals to the cell size. Therefore, a large number of cells are needed if a high localization accuracy is required, resulting in significantly reduced OWC capacity. To overcome this limit, we have reviewed the received signal strength-based accuracy improvement scheme in Section 6.3. By exploring the optical intensity distribution property and the received localization signal power, much better than cell size localization accuracy has been achieved. Demonstration of the localization accuracy has also been presented, and results have shown that an accuracy of better than 20 cm can be achieved with a cell size of 2 m. In addition, the accuracy limiting factors in the received signal strength-based scheme has been analyzed in this section. Results have shown that the major limiting factor is the background light. The advanced background light power estimation method to suppress its impact and to achieve better localization accuracy has also been introduced.

In the previous optical wireless localization schemes, it is assumed that the height information of the user transceiver is pre-known. However, it is not the case in most practical scenarios especially when the user is moving. To overcome this limit, we have introduced the 3D optical wireless localization principle capable of height estimation in Section 6.4. The 3D location of user transceiver has been realized by jointly exploring the received signal strength and the signal angle-of-arrival information, and the angle-of-arrival information is enabled by using the single-channel imaging receiver. Results have shown that better than 10 cm localization can be achieved for both height and user location estimation.

It is also worth noting that in addition to the near-infrared indoor optical wireless localization schemes discussed in this chapter, the visible range LEDs have also been explored to provide user location information. The signal time-of-arrival, strength and angle-of-arrival information have all been used. Interested readers can refer to references [20–24] for more details.

References

1. F. Lassabe, P. Canalda, Chatonnay, and F. Spies, Indoor WiFi positioning: Techniques and systems. *Annals of Telecommunications*, 2009. **64**(9–10): pp. 651–664.
2. S. A. Golden and S. S. Bateman, Sensor measurements for WiFi location with emphasis on time-of-arrival ranging. *IEEE Transactions on Mobile Computing*, 2007. **6**(10): pp. 1185–1198.
3. Y. Zhao, L. Dong, J. Wang, B. Hu, and Y. Fu, Implementing indoor positioning system via Zigbee devices, in *Asilomar Conference on Signals, Systems and Computers*, Pacific Grove, CA, 2008.
4. W. H. Kuo, Y. S. Chen, G. T. Jen, and T. W. Lu, An intelligent positioning approach: RSSI-based indoor and outdoor localization system scheme in Zigbee networks, in *International Conference of Machine Learning and Cybernetics (ICMLC)*, Qingdao, China, 2010.
5. B. A. Pahlavan and K. Pahlavan, Modeling of the TOA-based distance measurement error using UWB indoor radio measurements. *IEEE Communications Letters*, 2006. **10**(4): pp. 275–277.
6. C. Zhang, M. J. Kuhn, B. C. Merkl, A. E. Fathy, and M. R. Mahfouz, Real-time noncoherent UWB positioning radar with millimeter range accuracy: Theory and experiment. *IEEE Transactions on Microwave Theory and Technology*, 2010. **58**(1): pp. 9–20.
7. S. S. Saab and. Z. S. Nakad, A standalone RFID indoor positioning system using passive tags. *IEEE Transactions on Industrial Electronics*, 2011. **58**(5): pp. 1961–1970.
8. R. C. Luo and O. Chen, Wireless and pyroelectric sensory fusion system for indoor human/robot localization and monitoring. *IEEE/ASME Transactions on Mechatronics*, 2013. **18**(3): pp. 845–853.

9. T. K. Kohoutek, R. Mautz, and A. Donaubauer. Real-time indoor positioning using range imaging sensors, in *SPIE Photonics Europe*, Brussels, Belgium, 2010.

10. F. Walch, C. Hazirbas, L. Leal-Taixe, T. Sattler, S. Hilsenbeck, and D. Cremers. Image-based localization using LSTMs for structured feature correlation, in *IEEE International Conference on Computer Vision (ICCV)*, Venice, Italy, 2017.

11. M. S. Rahman, M. M. Haque, and K. D. Kim. High precision indoor positioning using lighting LED and image sensor, in *International Conference of Computer and Information Technology (ICCIT)*, Dhaka, Bangladesh, 2012.

12. S. Sayeef, U. K. Madawala, P. G. Handley, and D. Santoso. Indoor personal tracking using infrared beam scanning, in *Position Location and Navigation Symposium (PLANS)*, 2004.

13. K. Wang, A. Nirmalathas, C. Lim, and E. Skafidas. High speed 4× 12.5 Gbps WDM optical wireless communication systems for indoor applications, in *Optical Fiber Communication Conference*, OSA Publishing, Washington, DC, 2011.

14. K. Wang, A. Nirmalathas., C. Lim, and E. Skafidas, Experimental demonstration of a full-duplex indoor optical wireless communication system. *IEEE Photonics Technology Letters*, 2012. **24**(3): pp. 188–190.

15. K. Wang, A. Nirmalathas, C. Lim, and E. Skafidas, Experimental demonstration of a novel indoor optical wireless localization system for high-speed personal area networks. *Optics Letters*, 2015. **40**(7): pp. 1246–1249.

16. K. Wang, A. Nirmalathas, C. Lim, K. Alameh, H. Li, and E. Skafidas, Indoor infrared optical wireless localization system with background light power estimation capability. *Optics Express*, 2017. **25**(19): pp. 22923–22931.

17. K. Wang, A. Nirmalathas, C. Lim, K. Alameh, and E. Skafidas, Optical wireless-based indoor localization system employing a single-channel imaging receiver. *Journal of Lightwave Technology*, 2016. **34**(4): pp. 1141–1149.

18. M. Castillo-Vazquez and A. Puerta-Notario, Single-channel imaging receiver for optical wireless communications. *IEEE Communications Letters*, 2005. **9**(10): pp. 897–899.

19. K. Wang, A. Nirmalathas, C. Lim, and E. Skafidas, High-speed indoor optical wireless communication system with single-channel imaging receiver, *Optics Express*, 2012. **20**(8): pp. 8442–8456.

20. S.-Y. Jung, S. Hann, and C.-S. Park, TDOA-based optical wireless indoor localization using LED ceiling lamps. *IEEE Transactions on Consumer Electronics*, 2011. **57**(4): pp. 1592–1597.

21. U. Nadeem, N. U. Hasan, M. A. Pasha, and C. Yuen, Highly accurate 3D wireless indoor positioning system using white LED lights. *Electronics Letters*, 2014. **50**(11): pp. 828–830.

22. F. Mousa, N. Almaadeed, K. Busawon, A. Bouridane, R. Binns, and I. Elliott, Indoor visible light communication localization system utilizing received signal strength indication technique and trilateration method. *Optical Engineering*, 2018. **57**(1): p. 016107.

23. W. Guan, Y. Wu, S. Wen, et al., A novel three-dimensional indoor positioning algorithm design based on visible light communication. *Optics Communications*, 2017. **392**(1): pp. 282–293.

24. A. Arafa, S. Dalmiya, R. Klukas, and J. F. Holzman, Angle-of-arrival reception for optical wireless location technology. *Optics Express*, 2015. **23**(6): pp. 7755–7766.

7

Conclusions and Future Directions

7.1 Conclusions

In this book, we have focused on the use of the near-infrared optical wireless technology for high-speed wireless communications in indoor environments, especially personal working and living spaces. Compared to traditional RF wireless communications, the OWC systems have a much broader license-free bandwidth available and they are immune to electromagnetic interferences. The OWC link is also assumed to be highly secure, since it is challenging to intercept the light in the free space. Due to these unique advantages, the OWC technology has been considered as a promising candidate of providing ultra-high-speed wireless communications in the future in indoor environments to satisfy the rapidly growing communication speed requirement of end users and emerging applications.

To better understand the general principle and the capabilities, in Chapter 2 we have established the general theoretical model of near-infrared indoor OWC systems. The transmitter, the receiver and the OWC link have been first modeled separately, and both the direct LOS and the diffusive optical wireless links have been discussed. The complete system has then been modeled by combining different building blocks. The general modeling principle has been applied to a high-speed indoor OWC system example, which combines limited signal coverage with user tracking to enable user mobility, and the performance of system has been analyzed. In addition, the key limiting factor in indoor OWC systems, which is the background light always existing in indoor environments due to both sunlight and illumination lamps, has been analyzed. It has been shown that the inevitable background light results in higher power penalty when the data rate of system is lower.

In near-infrared indoor OWC systems, the LOS link needs to be used to realize high-speed operation. Therefore, this type of system is vulnerable to the physical shadowing problem, which cause the link blockage and service interruptions. To improve the robustness against the physical shadowing problem in high-speed indoor OWC systems we have introduced the spatial diversity principle in Chapter 3. Since the receiver is dedicated to each user, the receiver diversity usually results in relatively complicated receiver structure and associated cost. Therefore, we have focused on the transmitter diversity principle in this chapter. We have introduced two widely used transmitter diversity coding schemes RC and STBC, and we have described the basic operation principles. We have applied the system model established in Chapter 2 to analyze the performance of RC- and STBC-based indoor OWC systems. It has been shown that due to the use of direct detection, the RC scheme outperforms the STBC scheme. In indoor OWC systems with transmitter spatial diversity, the synchronization of multiple optical wireless channels is required, which can be challenging in practical applications. Therefore, we have also introduced a delay-tolerant transmitter diversity principle in Chapter 3. In this scheme, orthogonal filters have been applied in multiple wireless channels and the impact of channel delay has been suppressed by using matched filters at the receiver side. Results have shown that with up to 10 symbol period channel delay, the power penalty is suppressed to be negligible.

In Chapter 4, we have further discussed the WDM technology to increase the aggregate wireless transmission data rate in near-infrared indoor OWC systems. By using multiple optical carriers in the same OWC link, the data rate has been significantly improve. The multi-user access principles to support multiple users simultaneously have also been introduced in this chapter. Both traditional multi-user access techniques, including FDMA, TDMA and CDMA, and the recently proposed TSC scheme have been reviewed. The performance of TSC scheme has been analyzed in more detail, especially against the imperfect timing issue that is highly like to exist in practical systems. Results have shown that the TSC scheme is robust against the imperfect timing. In addition, the advanced adaptive loading technique in the TSC-based multi-user system, where users with different channel conditions are served with data rates adaptively, has been introduced to further increase the effective system data rate and throughput.

In previous development and demonstration of indoor OWC systems, discrete components and devices are utilized, which results in relatively bulky systems, instable performance and high system complexity and cost. To solve these issues, we have discussed the photonic integration of near-infrared indoor OWC systems in Chapter 5. We have first reviewed the photonic integration technology in general, and then focused on the silicon photonic integration platform due to its advantages, especially the compatibility with current highly advanced CMOS technology. We have briefly reviewed commonly used both passive and active devices in silicon photonic integrated circuits, before further discussing the unique beam steering device for OWC systems. The silicon integrated optical phased array has been used to realize the beam steering function due to the planar structure, the stable performance and the compatibility with other photonic integrated devices. We have shown the general operation principle and the key building blocks of the silicon integrated beam steering device, and we have also presented a demonstration example of a high-speed near-infrared indoor OWC system based on a silicon photonic integrated circuit. The BER performance of the system has been characterized, and results have shown the capability and great potential of realizing high-speed indoor OWC transceivers and systems in the integrated version.

Generally, in high-speed near-infrared indoor OWC systems, because of the limited transmission power due to the laser safety regulation, only limited signal coverage can be provided, and the user localization function is required to provide user mobility. Therefore, we have discussed indoor localization technologies in Chapter 6. The currently available RF-based, pyroelectric sensor-based, image sensor-based and passive optical beams-based schemes have been reviewed and the major limitations have been analyzed. One key disadvantage of using these localization techniques in indoor OWC systems is that additional devices are required, leading to increased transceiver complexity and cost. Therefore, we have introduced the optical wireless-based indoor user localization principle, which is based on the "Search and Scan" process. By sending out the unique "Cell Number" code and receiving feedback from the user, the localization function has been achieved. To further improve the localization accuracy, the received signal strength information has been further explored, and results have shown that an average localization accuracy of about 15 cm can be achieved. The localization accuracy limiting factor has been analyzed, and it has been found that the majority of localization error is because of the impact of background light. To reduce this impact, an advanced indoor optical wireless localization principle with background light estimation capability has been introduced, and the location estimation accuracy has been improved significantly. In addition to locating the user transceiver in two dimensions, the height information is also highly demanded. To satisfy this need, we have discussed a 3D indoor optical wireless localization principle. The 3D localization with height estimation capability has been achieved by jointly exploring the received signal strength and the signal angle-of-arrival information, where the angle-of-arrival information has been enabled by the single-channel imaging receiver. Accurate location of the user has been realized using this principle.

7.2 Optical Interconnect Application of Near-Infrared Optical Wireless Technology

In this book, we have mainly focused on the application of near-infrared optical wireless technology in realizing high-speed wireless communications in indoor environments. In addition to this application, there are other potential applications of the technology, such as realizing high-speed optical interconnects in data centers and high-performance computing platforms. In this section, we will provide the general overview of this application to potentially inspire some interests amongst the readers.

The volume of data processed by data centers and high-performance computing platforms has increased explosively during the past a few years, and high-speed interconnects are highly demanded for the efficient data retrieval, linkage, transfer, processing and computation. The widely used electronic interconnects, such as transmission lines in printed circuit boards (PCBs) and electrical cables in data centers, face a number of fundamental challenges under high-speed operations, which mainly include the limited signal bandwidth, high transmission loss, long latency, severe signal interference and large heat dissipations. To overcome these limitations, the use of optical interconnects has been proposed and investigated for data centers and high-performance computing platforms [1].

According to the data transmission distance, as shown by Figure 1.11, interconnects can be roughly divided into three categories: the long-range intra-data-center or rack-to-rack interconnects, the medium-range card-to-card interconnects and the short-range chip-to-chip or on-chip interconnects. A number of optical interconnect solutions have been studied targeting at different categories of interconnects. For the long-range interconnects, the single-mode optical fiber together with high-order modulation techniques and advanced signal processing algorithms has been used [2]. On the other hand, the silicon photonic integration technology discussed in Chapter 5 has been investigated and demonstrated for short-range chip-to-chip and on-chip applications [3].

In addition to the short- and long-range optical interconnects, high-speed medium-range card-to-card optical interconnects are also highly desired to transfer data between electronic cards within the same rack, which are commonly based on PCBs. MMF ribbons and polymer waveguides-based optical interconnect schemes have been proposed and studied [4,5]. The general structures are shown in Figure 1.12 and Tb/s scale data transmission has been demonstrated using these schemes. However, these optical interconnect schemes use point-to-point data transmission links. Therefore, high-speed opto-electronic-opto (OEO) converters are required when the data needs to be transferred to other destinations, whilst such OEO converters are fundamentally challenging to realize and they are also normally power hungry.

To solve these issues, the use of free space as the data interconnection link has been studied, where the OWC technology provides high-speed data transmissions with flexibility [6,7]. The general principle has been briefly introduced in Section 1.4 and the basic structure of OWC-based card-to-card optical interconnects has been shown in Figure 1.13. In this scheme, data-carrying optical signals propagate via the free space directly from the transmitting electronic card to the receiving card. A link selection block is inserted at the transmitter side and it can steer the optical signal beam to different receivers according to the requirement dynamically. The link selection block in OWC-based optical interconnects has been realized using various technologies, including both signal refraction-based (e.g. liquid-crystal-on-silicon devices and opto-VLSI processors) [6] and signal reflection-based (e.g. MEMS steering mirrors) schemes [7]. With the link selection capability, the data switching function and flexible interconnects are realized without requiring high-speed OEO converters.

In optical interconnects, since they are mainly used inside data centers and high-performance computing platforms, they are usually required in large volume. Therefore, low-complexity transceivers are needed for low cost considerations. As a result, VCSELs are widely employed as light sources and the direct modulation is used in OWC-based card-to-card optical interconnects to further reduce the system cost and complexity. However, the modulation bandwidth of VCSELs is limited, which leads to the relatively low interconnection speed compared to other types of optical interconnects, especially those using more advanced lasers as light sources.

To improve the interconnect speed in the VCSELs-based free-space optical interconnects, multiplexing techniques can be utilized. The WDM technology, which transmits multiple data-carrying optical signals with different wavelengths through the same OWC link, is one option. Whilst the interconnect speed can be significantly increased by using a large number of wavelength channels, the implementation using VCSELs is difficult, since the VCSEL normally has a broad linewidth and it is challenging to precisely control the radiation wavelength. Therefore, the spatial multiplexing scheme, where multiple optical interconnect links are operated simultaneously in parallel, is widely adopted. Although the speed of each channel is still limited, high aggregate data rates can be achieved. The channel spacing between neighboring parallel channels is a key design parameter in this type of optical interconnects to avoid significant inter-channel crosstalk.

In addition to the low modulation speed, the maximum radiation power from VCSELs is normally limited as well, and it leads to the power budget issue in optical interconnects. The power budget issue becomes even worse when considering the inherent optical beam divergence while propagating through the free space, where the optical signal beam footprint becomes larger at the receiver side. Therefore, the signal power that can be collected after interconnection is small and the achievable interconnection range is generally limited. The inter-channel crosstalk when multiple parallel OWC channels are used simultaneously to increase the aggregate interconnect speed also limits the interconnect range. Due to these reasons, the OWC-based optical interconnect range in previously reported demonstrations is only a few tens of centimeters [8].

To overcome the speed and interconnect range limitations described above, here we introduce a high-speed carrierless-amplitude-phase (CAP) modulated reconfigurable free-space optical interconnects for board-to-board applications employing STBC. The CAP modulation can be realized with low cost transceiver and it can directly modulate VCSELs. At the same time, it increases the data rate by enabling higher spectral efficiency. On the other hand, the STBC, which has been introduced in Chapter 3 as an effective transmitter diversity technique, is capable of extending the interconnect range through collecting larger signal power and reducing the impact of inter-channel crosstalk.

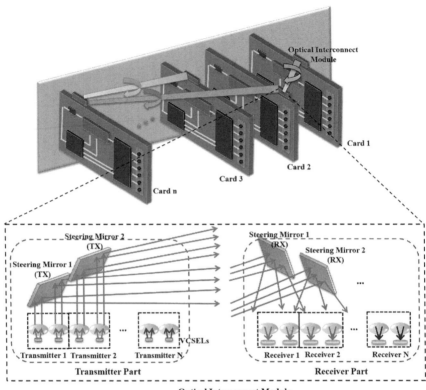

FIGURE 7.1
OWC-based optical interconnects with CAP and STBC.

The architecture of the high-speed free-space optical interconnects with extended range through CAP and STBC is shown in Figure 7.1. Similar with previously studied free-space optical interconnects, a dedicated optical interconnect module is integrated together with each electronic board (i.e. PCB) in the data center or the high-performance computing rack. Both OWC transmitter and receiver are integrated in the same interconnect module. The transmitter mainly consists of the VCSEL array, the micro-lens array and the MEMS mirror array. MEMS mirrors are used here as the link selection block to provide the OWC link reconfiguration capability. Inside the optical interconnect transmitter, VCSELs serve as light sources, and the generated optical beams pass through a micro-lens array for collimation. Then MEMS steering mirrors dynamically change beam directions towards required destinations.

After free-space propagation, data-carrying optical beams arrive at the receiver of the optical interconnect module, which consists of the MEMS mirror array, the micro-lens array and the photodiode (PD) array. MEMS mirrors are also used at the receiver to guide signal beams towards corresponding receiving elements, where a micro-lens array is used for collecting and focusing signals beams and a PD array is placed on the focal plane of micro-lenses for signal detection. The converted electrical signals are then processed to complete data interconnections.

As discussed above, the speed of such OWC-based optical interconnects is mainly limited by the modulation bandwidth of low-cost VCSLEs, and the range is usually limited by the output optical power of low-cost VCSELs, the Gaussian beam divergence while propagating through the free space and the inter-channel crosstalk. To extend the achievable optical interconnect range and to improve the interconnect speed, the CAP modulated and space-time coded free-space optical interconnects scheme can be used. The CAP and STBC transmitter and receiver structures are shown in Figure 7.2a and b, respectively. At the transmitter side, the binary data to be transmitted is first split into two streams a_n and b_n. The two streams then pass through two CAP pulse shaping filters, namely, in-phase and quadrature filters, respectively. The filters are based on the square-root raised cosine shape and they are orthogonal and form a Hilbert pair [9]. The two streams of signals, denoted as $s_{CAP, \text{In-Phase}}$ and $s_{CAP, \text{Quadrature}}$, are then summed up to generate the CAP symbol, which can be expressed as:

$$s_{CAP} = s_{CAP, \text{In-Phase}} + s_{CAP, \text{Quadrature}} \tag{7.1}$$

After CAP modulation, the generated symbols are then coded with STBC. The extended Alamouti-type STBC is employed here. Due to the non-negativity

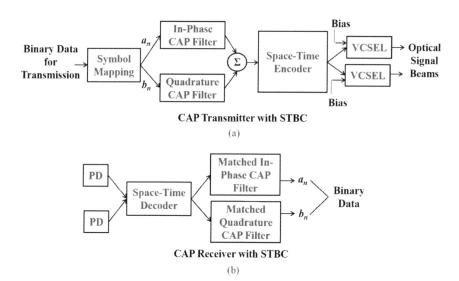

CAP Transmitter with STBC

(a)

CAP Receiver with STBC

(b)

FIGURE 7.2
(a) Principle of CAP and STBC transmitter and (b) principle of CAP and STBC receiver.

property, this type of STBC is suitable for implementation in optical systems with direct modulation and direct detection. For simplicity, here we focus on the case of using 2×2 STBC in the optical interconnect. Larger-scale STBC can be implemented using the similar approach. To implement the 2×2 STBC, two VCSELs and two PDs in each optical interconnect module form a transmitting element and a receiving element, respectively, which is shown in the inset of Figure 7.1. According to the working principle of STBC discussed in Chapter 3, each symbol transmission occupies two adjacent time slots, and the transmission matrix can be described as:

$$S = \begin{bmatrix} s_{CAP,1} & s_{CAP,2} \\ \sim s_{CAP,2} & s_{CAP,1} \end{bmatrix} \tag{7.2}$$

where $s_{CAP,1}$ and $s_{CAP,2}$ are the CAP symbols in the first and the second time slots, and $\sim s$ represents the bitwise not operation (i.e. $\sim s = 1 - s$). The rows of the transmission matrix correspond to the time slot k ($k = 1$ or 2) and the columns represent the VCSEL i ($i = 1$ or 2) in each transmitting element. After the OWC link propagation, the signals from two VCSELs in each transmitting element are collected and detected by the PDs in each receiving element. The signal received by PD j ($j = 1$ or 2) during time slot k can be expressed as:

$$y_{i,k} = \sum_{i=1}^{2} h_{j,i,k} s_{i,k} + N_{j,k} \tag{7.3}$$

where $h_{j,i,k}$ is the channel impulse response from the ith VCSEL to the jth PD during the time slot k and $N_{j,k}$ is the additive noise received by PD j. In optical interconnect applications, since the working environment is relatively stable, the OWC channels can be considered as slow-varying, and hence, $h_{j,i,1} \approx h_{j,i,2}$. The additive noise in optical interconnects can be considered as constant during two time slots as well (i.e., $N_{j,1} \approx N_{j,2}$). At the receiver side, after photo-detection by two PDs in each receiving element, the signals are first recovered from the STBC via the space-time decoder using the channel response information. Then CAP symbols are demodulated by passing through matched in-phase and quadrature pulse shaping filters. The two data streams are then mapped back to the original binary signal and the high-speed interconnect function is realized.

In the CAP- and STBC-based free-space optical interconnects, the CAP scheme enables more spectral efficient signal modulation directly onto VCSELs, and the implementation can be realized using appropriately designed electrical filters. Therefore, both high-speed data transmission and low complexity can be achieved simultaneously. On the other hand, the extended Alamouti-type STBC is employed in the optical interconnect to extend the data transmission range. This is achieved through two major aspects. First, two VCSELs are used in each transmitting element, and the signals transmitted by both VCSELs are collected and detected by the receiving element. Therefore, the total received signal power is larger (almost doubled) compared to previous OWC-based optical interconnects, where the received signal power comes from only one VCSEL and the interconnect range can be extended. Second, since the channel responses in the STBC-based optical interconnect system are obtained through the channel training process, the inter-channel crosstalk between VCSELs in each transmitting element can be turned into useful information for detection and decoding. Therefore, the inter-channel crosstalk in optical interconnects is effectively suppressed and the receiver sensitivity is improved, which can further extend the interconnect range.

Using the CAP and STBC principles in OWC-based optical interconnects, 40 Gb/s per channel interconnect speed has been experimentally demonstrated, and the achievable interconnect range has been extended by about 65% when two VCSELs are used inside each transmitting element [10]. Due to the suppressed inter-channel crosstalk as described above, the receiver sensitivity has been improved by about 1 dB. To further extend the optical interconnect range, higher-order STBC can be used, where a larger number of VCSELs are used in each transmitting element to increase the aggregate signal power available.

The optical interconnect application discussed in this section is one example of other possible application areas of the near-infrared optical wireless technology. Similar with indoor OWC systems, the optical interconnect also explores the inherent advantages of using the optical signal as the data carrier, which mainly include the broad bandwidth and the low transmission loss. Although exciting results have been achieved, there are still many aspects that require further study and investigation to enable practical applications. We will discuss some of the future research and development directions in the next section.

7.3 Future Research Needs and Directions

The near-infrared optical wireless technology has witnessed intensive interests during the past a few years and a large number of promising results have been achieved. However, there are still many fundamental and practical issues that remain unsolved, which hinder practical applications of this technology. We will end this book by discussing some of these issues to highlight future research needs and possible future research directions. We limit our discussion to the near-infrared high-speed indoor OWC system, which is the focus of the book.

1. Large-scale MIMO in indoor OWC systems: in Chapter 3, we have introduced the MIMO principle in indoor OWC systems, which explores the spatial diversity advantage. Up to now, most research focuses on small-scale transmitter and receiver diversity principles,

and the major reasons include the associated transceiver complexity and the control challenge of multiple transceivers. Therefore, the spatial diversity is mostly studied in small indoor environments with only few users, such as a small office scenario. For larger indoor space applications with more users, such as the shopping center, larger-scale MIMO is highly needed and it requires further study. The silicon photonic integration technology discussed in Chapter 5 is a promising solution to solve the transceiver complexity issue when large-scale transmitter and receiver diversities are used. It can integrate a large number of optical devices and functions, such as multiple modulators, optical filters, beam steering devices and optical detectors, onto the same small-size circuit. All the devices and functions of the photonic integrated circuit can be fabricated simultaneously, and hence, low-cost potentially can be achieved as well. To realize this promising solution, many aspects still require further investigation, such as the layout of multiple transmitters and receivers in the integrated circuit. In addition, with a large number of OWC transceivers available, the control, coordination and operation of these spatial transceivers also require further fundamental study. Large-scale MIMO has been studied in RF systems. However, the MIMO principle is fundamentally different in OWC systems, due to different channel characteristics and the typical use of direct detection. Therefore, the theoretical modeling of large-scale MIMO in OWC systems is required. Due to the unique channel property, the layout of multiple OWC transceivers in indoor environments also requires further study and optimization. The multiple OWC transceivers can be operated in either coordinated or non-coordinated methods, and the fundamental capabilities and limitations of both methods also need to be analyzed both theoretically and experimentally. When the coordinated MIMO scheme is employed, the control of multiple transceivers, such as the synchronization and the signal distribution, need to be studied as well.

2. Robust high-speed indoor OWC systems: currently, the LOS link is used to realize high-speed wireless communications in near-infrared indoor OWC systems. However, such link configuration is vulnerable to physical shadowing, which is highly likely when the user is moving. The use of spatial diversity has been introduced in Chapter 3 to improve the system robustness by providing redundant OWC links. However, simultaneous blockage of multiple channels is still possible, and the chance is high when the indoor environment is relatively complicated, such as in offices with many cubicles and furniture. Therefore, realizing robust indoor OWC system with high communication speed is still challenging and further research is necessary. One possible solution is combining the LOS and the diffusive

links in the same indoor OWC systems. Due to the use of reflected and diffracted signals, the diffusive link can provide full coverage over the entire indoor environment by optimizing the locations of light sources. Therefore, the wireless communication is robust. However, as discussed in Chapter 2, the channel bandwidth is highly limited to large multipath dispersion. As a result, the diffusive link needs to be combined with the direct LOS link to achieve both high-speed and robustness. The combining method and the switching between two types of links need to be systematically studied. In addition, channel bandwidth improvement method of diffusive link-based indoor OWC systems also requires further research.

3. Multi-user access in indoor OWC systems: in practical indoor OWC systems, multiple users and communication devices need to be served simultaneously. We have discussed some general multi-user access principles in Chapter 4, including the traditional FDMA, TDMA and CDMA, and the TSC schemes. However, one physical requirement is that the users served using these principles need to be located inside the same signal beam coverage area, due to the limited transmission power. When multiple users located further apart need to be served simultaneously, physical beam steering is needed and the multi-user access principles mentioned above cannot be used. Therefore, the corresponding multi-user access technique needs to be further studied. When the transmitter diversity principle is used, multiple transceivers potentially can be coordinated to serve different users simultaneously. In this case, the control and the coordination of multiple transmitters require detailed investigations. The physical limit of beam steering devices, such as the switching time, further complicates the scenario, and further study is needed.

4. Miniaturization of transceivers in indoor OWC systems: as discussed in Chapter 5, current indoor OWC systems primarily rely on discrete components and the entire system is normally bulky. This is especially problematic for the user transceiver, since it is directly located together with moving users. Take the receiver part as an example, which mainly consists of the optical beam concentrator and the PD. The optical concentrator, especially non-imaging concentrator, normally has large size, and hence, limiting the profile of the user transceiver. This problem becomes even worse when advanced receivers are used, such as the angle diversity receiver with multiple optical concentrators. To reduce the size of indoor OWC systems and transceivers, the silicon photonic integration platform is a promising solution. However, current studies of silicon photonics mainly focus on waveguide-type of applications, and the free-space interfaces for OWC systems need to be further investigated. For example,

the optical phased array can be used for collecting the optical signal after free-space propagation in principle. However, very limited research has been conducted, which leaves an open question.

5. Uplink technologies in indoor OWC systems: the technologies and principles discussed in this book focus on the downlink in indoor OWC systems, which transmits high-speed data to the end users. In addition to the downlink, the uplink, which transmits data generated by the end users, is also a necessity. Compared to the downlink, the speed requirement of the uplink communication is lower, and hence, it is easier to realize in principle. However, the uplink OWC faces even more challenges in reality, which are mainly introduced by resource and cost limitations. For example, since the uplink transmitter is located at the moving user side, relatively simple structure, low-cost and high stability are required. In addition, the uplink link transmitter is highly likely to be operated using battery power, and hence, low power consumption is necessary. The uplink transmitter is also more likely to be blocked compared to the downlink transmitter, since it is closer to the end user. Therefore, the robustness requirement is even more stringent. Due to these and other unmentioned constrains, practical indoor OWC uplink requires significant further research attention. Furthermore, when the downlink transmitter diversity principle is implemented, it also means that the uplink system has multiple receivers available. Therefore, the efficient use of these receiver resources to realize robust uplink with low power consumption and moderate speed is also an interesting area that needs to be further explored.

6. Beam shaping techniques for more efficient signal distributions in indoor OWC systems: due to the maximum transmission power limited by the laser safety regulation, only limited signal coverage can be provided in high-speed near-infrared indoor OWC systems. The optical power distribution inside the signal coverage area normally can be seen as approximately Gaussian. Therefore, the optical intensity is higher at locations that are closer to the beam center, and it decreases when moving further away from the beam center. In order to provide high-speed wireless connectivity inside the entire beam coverage area, the power available at the beam edges needs to be higher than the receiver sensitivity. In this case, the optical power that can be collected around the beam center is higher than necessary. In another word, the optical power at the receiver side is not efficiently utilized. If the optical power is more evenly distributed inside the signal beam coverage area, a smaller transmission power can be used when the coverage area is kept constant, or a larger coverage area can be realized when

the transmission power is kept unchanged. To change the intensity distribution, the beam shaping is needed in indoor OWC systems, which has received very limited attention so far. Low insertion loss and broad bandwidth are two key requirements of the beam shaping in indoor OWC systems, to avoid significant power attenuation and to allow the transmission of high-speed signals. The shaped beam also needs to have a controllable divergence, to realize a pre-designed beam coverage area. The complexity of the beam shaping scheme needs to be relatively low as well, and it is highly desirable to be compatible with the photonic integration platform. Therefore, further study is highly required on possible beam shaping techniques.

7. Optical wireless communication networks: current studies focus more on the subsystem and system techniques of using the optical wireless principle to provide high-speed data transmissions. For practical applications, the networking of optical wireless transceivers, subsystems and systems is required, including both the physical layer and the media access control (MAC) layer. Here we take the MAC layer as an example. MAC layer protocols for RF-based wireless LANs have been widely studied and implemented. However, such protocols may not be suitable for indoor OWC networks. One of the main reasons is the large discrepancy in the communication speed. The typical speed in RF networks is in the order of several tens of Mbps to hundreds of Mbps, whilst it is several Gbps in OWC networks. Therefore, more efficient MAC layer protocols with higher throughput and lower latency are needed. In addition, the MAC protocols in indoor OWC networks also need to take the physical hardware limits into consideration, especially those imposed by the beam steering function, such as the microsecond scale switching time. Furthermore, the uplink MAC layer protocols require even more advanced functionalities, since the uplink transmissions by different users need to be appropriately coordinated to avoid data collision and transmission failure. All these aspects require further study.

The possible future research directions in indoor near-infrared OWC systems listed above are just some examples based on the authors' knowledge and experience. There are many more research needs in this exciting field. It is also worth mentioning that in addition to the near-infrared indoor OWC systems, VLC systems are also promising for providing high-speed wireless communications in indoor environments in the future. A large number of researches have been carried out on VLC systems, and interested readers are encouraged to refer to the book [11] for more details.

References

1. M.A. Taubenbaltt, Optical interconnects for high-performance computing. *Journal of Lightwave Technology*, 2012. **30**(4): pp. 448–457.
2. E. El-Fiky, M. Chagnon, M. Sowailem, A. Samani, M. Morsy-Osman, and D.V. Plant, 168-Gb/s single carrier PAM4 transmission for intra-data center optical interconnects. *IEEE Photonic Technology Letters*, 2017. **29**(3): pp. 314–317.
3. H. Subbaraman, X. Xu, A. Hosseini, X. Zhang, Y. Zhang, D. Kwong, and R.T. Chen, Recent advances in silicon-based passive and active optical interconnects. *Optics Express*, 2015. **23**(3): pp. 2487–2511.
4. R. Dangel, J. Hofrichter, F. Horst, D. Jubin, A. La Porta, N. Meier, I.M. Soganci, J. Weiss, and B.J. Offrein, Polymer waveguides for electro-optical integration in data centers and high-performance computers. *Optics Express*, 2015. **23**(4): pp. 4736–4750.
5. F.E. Doany, B.G. Lee, D.M. Kuchta, A.V. Rylyakov, C. Baks, C. Jahnes, F. Libsch, and C.L. Schow, Terabit/sec VCSEL-based 48-channel optical module based on holey CMOS transceiver IC. *Journal of Lightwave Technology*, 2013. **31**(4): pp. 672–680.
6. C.J. Henderson, D.G. Leyva, and T.D. Wilkinson, Free space adaptive optical interconnect at 1.25 Gb/s with beam steering using a ferroelectric liquid-crystal SLM. *Journal of Lightwave Technology*, 2006. **24**(5): pp. 1989–1997.
7. K. Wang, A. Nirmalathas, C. Lim, E. Skafidas, and K. Alameh, Experimental demonstration of 3 × 3 10 Gb/s reconfigurable free-space optical card-to-card interconnects. *Optics Letters*, 2012. **37**(13): pp. 2553–2555.
8. K. Wang, A. Nirmalathas, C. Lim, K. Alameh, and E. Skafidas, Experimental demonstration of high-speed free-space reconfigurable card-to-card optical interconnects. *Optics Express*, 2013. **21**(3): pp. 2850–2861.
9. K. Wang, A. Nirmalathas, C. Lim, E. Skafidas, and K. Alameh, Full-duplex gigabit indoor optical wireless communication system with CAP modulation. *IEEE Photonics Technology Letters*, 2016. **28**(7): pp. 790–793.
10. K. Wang, A. Nirmalathas, C. Lim, K. Alameh, and E. Skafidas, High-speed reconfigurable free-space optical interconnects with carrierless-amplitude-phase modulation and space-time-block code. *Journal of Lightwave Technology*, 2019. **37**(2): pp. 627–633.
11. S. Arnon, *Visible Light Communication*. Cambridge, UK: Cambridge University Press, 2015.

Index

Printed and bound by CPI Group (UK) Ltd, Croydon, CR0 4YY

17/10/2024

01775682-0009